THOMAS REID ON THE ANIMATE CREATION

The Edinburgh Edition of Thomas Reid

Series Editor: Knud Haakonssen

THOMAS REID ON THE ANIMATE CREATION

Papers Relating to the Life Sciences

EDITED BY
Paul Wood

THE PENNSYLVANIA STATE
UNIVERSITY PRESS
University Park, Pennsylvania

In memory of my parents

© Paul Wood, 1995

First published 1996 in the United States by
The Pennsylvania State University Press,
Suite C, 820 North University Drive,
University Park, PA 16802

All rights reserved

Typeset in 10½/13½pt Times
by Speedspools, Edinburgh, and
printed and bound in Great Britain

A CIP record for this book is available
from the Library of Congress

ISBN 0–271–01571–3

It is the policy of The Pennsylvania State University Press
to use acid-free paper for the first printing of all
clothbound books. Publications on uncoated stock satisfy
minimum requirements of American National Standard for Information
Sciences – Permanence of Paper for Printed Library Materials,
ANSI 239.48–1984

CONTENTS

Acknowledgements	vi
Abbreviations	vii
Editorial Principles	ix
Index of Manuscripts	xiii
INTRODUCTION	1
1 Natural History	3
2 Physiology	20
3 Materialism	30
Notes	56
THE MANUSCRIPTS	79
Part One Natural History	81
Part Two Physiology	99
Part Three Materialism	125
NOTES	243
Explanatory Notes	245
Textual Notes	257
Index	271

ACKNOWLEDGEMENTS

This volume has been longer in the making than I care to remember. During its protracted period of gestation I have incurred many debts. For financial support I am grateful to the University of Toronto, the Advisory Research Committee of Queen's University, and particularly to the Social Sciences and Humanities Research Council of Canada. The librarians of Aberdeen, Edinburgh and Glasgow University Libraries, the Andersonian Library (University of Strathclyde) and the Mitchell Library, Glasgow, have graciously allowed me to use manuscripts in their care, as has the Keeper of the Records of Scotland. I owe a special debt of thanks to Colin McLaren (Head of Special Collections and University Archivist, Aberdeen University Library) for his help with various projects; to Dorothy Johnston (now of Nottingham University) for sharing with me her detailed knowledge of the Birkwood Collection and related materials; and to Myrtle Anderson-Smith and Iain Beavan for their friendly assistance.

This volume has its distant origins in a plan to publish all of Thomas Reid's extant manuscripts, and I would like to thank Charles Stewart-Robertson for inviting me to contribute to that sadly abortive project. David Fate Norton has been a constant source of encouragement and sound advice over the years, as have Roger Emerson, and Knud Haakonssen, who knows all too well the practical difficulties involved in editing Reid's papers. Tom Cleary, Ian Douglas, Kurtis Kitagawa, John Money, Richard Sher, and John Wright provided me with valuable information and suggestions. I am especially grateful to M. A. Stewart for his translation of some Latin passages and for his help regarding editorial matters. I wish I could give adequate expression to what I owe to Carol and Hilary for their contribution to this book. Finally, I am sure that all of the individuals mentioned above would agree that I am solely responsible for any remaining errors.

ABBREVIATIONS

Libraries
AL   Andersonian Library, University of Strathclyde
AUL   Aberdeen University Library
EUL   Edinburgh University Library
GUL   Glasgow University Library
SRO   Scottish Record Office

Works by Thomas Reid

Active Powers   *Essays on the active powers of man* (Edinburgh, 1788)

Inquiry   *An inquiry into the human mind, on the principles of common sense*, 4th edn (London, 1785)

Intellectual Powers   *Essays on the intellectual powers of man* (Edinburgh, 1785)

'Notes'   George Baird, 'Notes from the Lectures of Dr Thomas Reid, 1779–80', 8 vols, The Mitchell Library Glasgow, MS A104929

Works   *The works of Thomas Reid, D.D. now fully collected, with selections from his unpublished letters*, ed. Sir William Hamilton, 4th edn (Edinburgh, 1854)

Works by Joseph Priestley

Priestley, *A free discussion*   *A free discussion of the doctrines of materialism, and philosophical necessity, in a correspondence between Dr. Price, and Dr. Priestley* (London, 1778)

Priestley, *Examination*   *An examination of Dr. Reid's inquiry into the human mind on the principles of common sense, Dr. Beattie's essay on the nature and immutability of truth, and Dr. Oswald's appeal to common sense in behalf of religion* (London, 1774)

Priestley, *Disquisitions*   *Disquisitions relating to matter and spirit* (London, 1777)

Priestley, *Hartley*   *Hartley's theory of the human mind, on the principle of the association of ideas; with essays relating to the subject of it* (London, 1775)

Priestley, *Institutes*   *Institutes of natural and revealed religion*, 3 vols (London, 1772–4)

Miscellaneous

Descartes   *The philosophical writings of Descartes*, trans. John Cottingham, Robert Stoothoff, Dugald Murdoch and Anthony Kenny, 3 vols (Cambridge, 1984–91)

Fraser   A. Campbell Fraser, *Thomas Reid* (Edinburgh and London, [1898])

McCosh   James McCosh, *The Scottish philosophy, biographical, expository, critical, from Hutcheson to Hamilton* (London, 1875)

Newton, *Opticks*   Sir Isaac Newton, *Opticks: or, a treatise on the reflections, refractions, inflections & colours of light*, 4th edn (London, 1730; New York, 1952)

Newton, *Principia*   *Sir Isaac Newton's mathematical principles of natural philosophy and his system of the world*, rev. trans. Florian Cajori, 2 vols (Berkeley, Los Angeles and London, 1974)

Phil. trans.   *Philosophical transactions of the Royal Society of London*

Stewart   Dugald Stewart, *Account of the life and writings of Thomas Reid, D.D. F.R.S.Edin. late Professor of Moral Philosophy in the University of Glasgow* (Edinburgh, 1803)

Wood   P. B. Wood, 'Thomas Reid, natural philosopher: a study of science and philosophy in the Scottish Enlightenment', unpublished Ph.D. dissertation, University of Leeds, 1984

EDITORIAL PRINCIPLES

My primary aim as an editor has been to provide accurate, accessible and easy-to-use texts. Rather than adopt an elaborate set of editorial conventions designed to capture all of the textual alterations within Reid's manuscripts, I have opted for a more pragmatic approach to the task of translating Reid's unpublished writings on to the printed page. I have done so in the firm belief that it is pointless to employ a highly complex editorial apparatus to recreate what is in effect a textual facsimile – scholars interested in this kind of detail are better served by photographic reproductions or consultation of the original materials. Since the turn of the nineteenth century, Reid's unpublished writings have largely been ignored by those who purport to study him, and I hope that, by presenting Reid's texts in a readable form, this volume (like Knud Haakonssen's edition of Reid's papers on natural jurisprudence) will illustrate the intellectual riches of the Reid materials in Aberdeen University Library, and thereby encourage the further exploration of his surviving manuscripts.

I should emphasize that the manuscripts transcribed below represent only a selection of those related to the topics covered in this volume. Of the 639 Reid items contained in the Birkwood Collection (AUL MSS 2131/1–8), in the AUL MS 3061 series, and in the papers of Thomas Gordon and Robert Eden Scott (AUL MS 3107/1–9) approximately 265 (42 per cent) deal with mathematics and the natural sciences. Given the size of the Reid archive, some editorial decisions had to be made regarding what manuscripts to publish. The mathematical and scientific papers (like the Reid collections more generally) can be divided into those manuscripts primarily related to teaching, and those connected with Reid's activities outside of the classroom; these, in turn, can be divided into manuscripts connected with his publications and his participation in the Aberdeen Philosophical and Glasgow Literary Societies, reading notes, letters, and writings on miscellaneous topics. I have excluded from this volume all manuscripts directly related to Reid's teaching, since these form a coherent group which deserve separate publication. Among the remainder of the manuscripts, I have largely excluded Reid's extant reading notes, except for those which illustrate important aspects of the development of Reid's thought. Nor have I incorporated any of the relevant unpublished correspondence in this volume, because a complete edition of Reid's correspondence is planned. Some of the papers transcribed below have their origins in the proceedings of the Glasgow Literary Society, and I have included them here because their contents are more readily understood in the context of Reid's unpublished work on related scientific subjects. Finally, I have collected together those manuscripts which do most to illuminate a crucial aspect of his career, but which has been virtually unstudied hitherto, namely, his engagement with the culture of the eighteenth-century life sciences.

In transcribing the manuscripts I have chosen for inclusion, I have adopted the following editorial conventions:

1. I have made no attempt to normalize Reid's erratic spelling and punctuation, except where the spelling was clearly mistaken or where its eccentricity was too distracting. While I recognize that some readers might prefer a thorough modernization of spelling and punctuation, I see no point in eliminating the original flavour of Reid's manuscripts. Reid is not our contemporary, and we should not delude ourselves by completely modernizing his texts.

2. I have restricted the use of square brackets to the Introduction, and there they are employed either within quotations to adjust a text to its context or in footnotes to supply editorial information.

3. In my transcriptions, words or characters which are missing because of damage to the manuscript or are judged to have been inadvertently omitted by Reid are enclosed thus, '⟨ ⟩'. I have silently normalized Reid's contractions and abbreviations where no modern equivalent exists, or where they are not self-explanatory or readily pronounceable by the modern reader. Thus, on p. 128, l. 21, I have not expanded 'Revd' because it is still recognized as an abbreviation for 'Reverend', but in ll. 21–2 I have normalized Reid's contraction 'ABishop' because it does not

obviously stand for 'Archbishop'. A far more extreme example of a contraction which I have silently normalized occurs in MS 2131/2/III/13, A recto, where Reid uses 'S I N' as an abbreviation for 'Sir Isaac Newton'. I have also silently expanded some of Reid's contractions in the interests of readability. For example, on p. 128, l. 28 I have normalized Reid's 'Prop' to 'Proposition' because it occurs in a reading context. But where contractions occur in the context of a reference (as on p. 128, l. 28), I have usually left them unexpanded because their meaning will become obvious from the information supplied in the explanatory notes.

Characters which Reid wrote as superscripts are here printed on the regular line. In his manuscripts, Reid normally overlines for emphasis, and the relevant passages have been reproduced in italics without editorial comment.

4. Individual manuscript numbers and folio/page numbers are printed in the margins. Folio/page breaks are indicated by a vertical line '|' in the text.

5. Variants in Reid's manuscripts are recorded in the textual notes, and these are keyed to the relevant texts using page and line numbers (that is, '128/25' in the textual notes refers to p. 128, l. 25). In these notes, editorial comment is in italics and the manuscript texts are in regular typeface. Words repeated and left undeleted by Reid have not been recorded, nor have catchwords, nor those instances where Reid has changed an unfinished text by superimposing a letter or word on top of what he had written originally, or revised a phrase in the course of his initial writing. I have also omitted those instances where Reid has merely gone back and corrected his spelling or grammar, or supplied a missing word or words. Variants are indicated in the following manner. In MS 2131/2/III/13, (p. 223, ll. 25–6 below), for example, Reid initially wrote 'There may be, and very probably are combinations', then struck out the phrase 'may be, and very probably are', and substituted 'are no doubt' over the line. Reid's change is recorded in the textual notes thus:

223/25-6 are no doubt] may be, and very probably are

Where there are variants of variants I have usually followed this method of indicating Reid's changes, but in cases where this was not practicable I have explained the textual alterations in the notes.

Cancelled passages have been identified and recorded in the notes. Reid often failed to replace deleted material with a new word or phrase. On p. 224, l. 9, for example, he first wrote 'part nor symptom', and subsequently deleted 'part nor', so that in the state in which he left his manuscript only 'symptom' remains. This change is recorded in the textual notes thus:

> 224/9   symptom] part nor symptom

Reid sometimes left his initial formulation of a passage uncancelled. For instance, on p. 156, l. 13, Reid had initially written the phrase 'may enter the Brain at one part or another', and then wrote over the line 'primarily affect one part of the brain or another', without cancelling his original formulation. This change is recorded in the textual notes thus:

> 156/13   primarily affect one part of the brain or another] may enter the Brain at one part or another (*uncancelled*)

As for words or phrases which Reid has added to a finished text, I have recorded these in the following manner. On p. 224, l. 4, Reid added the word 'active' to the phrase 'various Functions'. In this (and similar cases) I have recorded the addition thus:

> 224/4   active *added*

In ambiguous cases I have specified the addition using the normal convention. Where Reid has placed his addition in the margin of the page, I have noted the location of the addition in my annotation.

Where appropriate, I have included in the textual notes information concerning the physical state of the manuscript.

The explanatory notes preceeding the textual notes contain translations of Latin passages, and the details of papers and books Reid quotes from or refers to in his texts. Materials in the explanatory notes are keyed to Reid's texts using the same convention employed in the textual notes. Detailed commentary on the contents of the manuscripts has been confined to the editor's introduction.

The editorial conventions employed here are similar to, but not identical with, those adopted by Knud Haakonssen in his edition of Reid's papers on natural jurisprudence. However, the differences are minor, and readers familiar with the conventions used in one volume will be able to consult the other without difficulty.

INDEX OF MANUSCRIPTS

This index lists all of the Thomas Reid manuscripts reproduced in whole or in part below, including manuscripts quoted (but not merely cited) in the Introduction.

Birkwood Collection Manuscripts

MS NUMBER	TITLE	PAGES IN THIS VOLUME
2131/2/I/15	Untitled	233–40
2131/2/III/13	Untitled	217–32, 241
2131/3/I/17	11 Junii 1788 Philosophiæ Recentoris a Benedicto Stay	171–3
2131/3/I/25	18 June 1774 Read Institutes of Natural & Revealed Religion	127–32
2131/3/II/14	Histoire Naturelle Generale & Particuliere	83–7
2131/3/II/16	Read Contemplation de la Nature	93–8
2131/3/III/3	Untitled	13, 14, 61
2131/3/III/23	Untitled	154–60
2131/4/I/29	Lectures on the Culture of the Human Mind	8, 9, 15
2131/4/II/1	Pneumatology	34
2131/6/I/17	Minutes of a Philosophical Club	10, 20
2131/6/V/12	Experiments upon Seminal Liquors	87–91
2131/6/V/35	Untitled	91–2
2131/7/II/2	Of the Involuntary motions of Animals	101–3
2131/7/II/19	Untitled	101
2131/8/V/1	Scheme of a Course of Philosophy	63, 66
2131/8/VII	Untitled	12

Aberdeen University Library Manuscripts

3061/1/4	Some Observations On the Modern System of Materialism.	173–217

3061/2	Of Muscular Motion in the human Body	103–24
3061/9	Miscelaneous Reflections on Priestley's Account of Hartley's Theory of the Human Mind	132–54
3061/12	Disquisitions relating to matter & spirit J Priestly 1777	161
3061/13	Free discussion	161–4
3061/14	May 1788 Read Disquisitions Relating to Matter & Spirit by J Priestly Lond 1777	168–71
3061/23	Untitled	164–8
K.160	Natural Philosophy 1758	70, 73

Introduction

1. Natural History

When describing Thomas Reid's life as Minister at New Machar, Dugald Stewart noted that Reid's 'chief relaxations were gardening and botany, to both of which pursuits he retained his attachment even in old age'.[1] As one of Reid's closest associates at the end of his career, Stewart was well placed to observe Reid's predilection for botany and gardening at first hand, but his characterization of the nature of Reid's botanical and horticultural interests while at New Machar is open to question for a number of reasons. First, Stewart saw the New Machar years (1737–51) as those in which Reid increasingly turned away from the scientific studies of his youth in order to devote himself to epistemology, and hence downplayed the significance of Reid's scientific activity in this period.[2] Secondly, while at New Machar Reid was on friendly terms with the noted agricultural improver Sir Archibald Grant of Monymusk, and their friendship suggests that Reid's gardening and botanizing may have been motivated in part by the improving spirit. Reid's theoretical and practical efforts in the realm of agricultural improvement while at King's College Aberdeen indicate too that he was probably something of an improver on a minor scale at the manse in New Machar.[3] Finally, we know from the sketchy record of the Philosophical Club in Aberdeen to which Reid belonged in 1736 and 1737 that on at least one occasion the discussion turned to the role of immaterial agents in the physiology of plants and animals, and hence it is probable that Reid's natural historical investigations at New Machar were inspired to some extent by his religious beliefs.[4] Consequently, Stewart can be criticized for trivializing Reid's early botanical and horticultural interests, since Reid's papers indicate that his botanizing had profound philosophical ramifications which Stewart apparently preferred to ignore.

The earliest of Reid's surviving manuscripts devoted to natural history date from the 1750s, and are largely related to his lectures on the subject given at King's College beginning in 1753.[5] Reading notes taken while preparing his lectures survive, and they show that Reid consulted such texts as Réaumur's *Memoires pour servir à l'histoire des insectes*, Tournefort's *Botanical institutions*, and Du Hamel de Monceau's *La Physique des arbres*.[6] No complete set of notes from the course is extant, but a reasonably detailed idea of the structure and contents of his lectures can be gained from Reid's 'Scheme of a Course of Philosophy', which he drafted in 1752. Reid here allotted classes to the anatomy of plants and animals along with a further session on their comparative anatomy, and a surviving lecture outline indicates that he drew on the works of Malpighi and Grew for his discussion of the vegetable kingdom.[7] Reid lectured too on the physiology of plants, animals and humankind, with the overt aim of demonstrating that their vital functions were the result of the actions of immaterial causes.[8] Reid also provided his students with classificatory schemes for, and chemical analyses of, the three kingdoms of nature, drawn from the works of Da Costa, Hill, Kunckel, Linnaeus and Ray, as well as Réaumur and Tournefort.[9] From the two extant versions of the text of his introductory lecture, we can see as well that Reid emphasized the relation of natural history to manufactures and the arts and its revelation of God's benevolent design in the creation.[10]

While at King's, Reid also engaged in field-work with his close associate David Skene, and with the Marischal Professor of Natural Philosophy (1760–75), George Skene. Among the voluminous papers of David Skene there are notes on plants observed by Skene and Reid while riding and walking in the Aberdeen area, and in the Birkwood papers there is a small notebook which Reid may have used on local simpling expeditions with Skene.[11] Following Reid's departure for Glasgow in September 1764, the two of them continued to correspond on natural historical topics, ranging from generation to petrifaction.[12] George Skene was apparently interested in chemical analysis, for he collaborated with Reid in analysing spa waters as well as various natural substances with a view to their proper classification.[13]

The most challenging text in natural history which Reid read while at King's College was undoubtedly Buffon's *Histoire naturelle*. Reid probably first turned to the *Histoire* in order to glean information for his natural history lectures, but, as the notes from the *Histoire* reproduced below show, he was soon caught up in a critical confrontation with Buffon's controversial ideas about classification, the history of the earth, the theory of generation and the nature of humankind.

Reid and his fellow Aberdonians were committed to a classificatory approach to the study of humankind and the three kingdoms of nature, and it is thus not surprising that Reid took exception to

Buffon's attack on taxonomy in the 'Premier discours' of the *Histoire*. Whereas Buffon argued that classification schemes were the creation of the human mind and had no ontological basis in nature, Reid insisted that God had created distinct classes, genera and species, and had formed our minds so that we can discover these divisions. Far from being fictions (as Buffon had claimed), taxonomic categories were for Reid rooted in the 'common sense' of mankind, and their validity was guaranteed by God's providential dispensation. Furthermore, Reid stressed the utilitarian benefits of systematic classification, and he also believed that we derive aesthetic satisfaction as well as a kind of religious instruction from the taxonomic enterprise.[14]

Reid similarly objected to Buffon's account of mathematical truths in the 'Premier discours'. According to Buffon, all mathematical derivations are essentially tautologous, since the conclusions are logically equivalent to the suppositions and definitions from which they are derived. Reid's exclamation in his notes, 'Credat Judaeus Apella', signals his disagreement with Buffon's position. Moreover, Reid could not accept Buffon's assertion in the *Histoire* that mathematical truths are the exclusive creations of the human mind. Reid believed that mathematical concepts are formed through a process of mental abstraction which begins with the raw materials of sensory experience, and he later wrote to his kinsman James Gregory that mathematical objects are 'possible modifications of things which we dayly perceive by our senses'. Consequently, Reid maintained that mathematical concepts have some (albeit tenuous) empirical basis, and hence that these concepts are not simply 'Creature[s] of the Mind'. As for the application of mathematics to the various branches of natural philosophy, there is little question that Reid found Buffon's suggestion that mathematical reasoning could only be used legitimately in astronomy and optics unacceptable. Reid took Newton's *Principia* to be the true exemplar of correct scientific practice, and his brand of Newtonianism led him to introduce quantitative methods into the highly qualitative field of chemistry. Reid's scientific ideals presupposed the validity of applying mathematics to the study of all natural phenomena, and it was this assumption which would have made him hostile to Buffon's overly restrictive view of the potential scope for mathematics in the physical sciences.[15]

Having read the 'Premier discours' with little satisfaction, Reid cannot have found much comfort in the following sections of the *Histoire naturelle* in which Buffon outlined his contentious cosmogonical and geological speculations. Although his notes do not betray any overt hostility to Buffon's system, it is clear from a passage in his natural theology lectures at Glasgow that Reid regarded this part of the *Histoire* as being in the tradition of world-makers like Burnet, Whiston and Woodward, and that he considered Buffon's theory of the earth to be a mere hypothesis. Reid no doubt stressed the conjectural status of Buffon's system because he wanted to discredit a manifestly heterodox explanation of the evolution of the earth. Buffon's French critics had already observed that he apparently assumed that the earth was much older than was commonly supposed, and Reid too seems to have perceived this, for he recorded in his reading notes that Buffon affirmed 'there are in many places even very high Mountains such vast collections of Marine substances that they cannot be supposed to be gathered together but in a long Course of years or perhaps Ages'. Even if he did not reflect on the implications of this particular claim, it is unlikely that Reid missed the significance of Buffon's refusal to invoke the Deluge as a geological cause and his insistence that we ought to explain the past in terms of the causes regularly operating in the present. For Buffon's methodological strategy was blatantly naturalistic, and brought into question the religious presuppositions which had informed most cosmogonical theorizing in the late seventeenth and early eighteenth centuries.[16]

Buffon's theory of generation must have struck Reid as subversively heterodox as well. In the *Histoire* Buffon asserted that organic matter is perpetually active and capable of organizing itself into plants and animals. To this 'matière vivante', as he called it, Buffon attributed a power analogous to that of gravity which effected the processes of nutrition, growth and reproduction. Furthermore, Buffon explicitly denied that God intervened in generation. Reid's reading notes show that he understood Buffon as claiming that organic molecules '*by a natural Power or Virtue* make various Conjunctions which seem to have life & Motion & perhaps really have so, these may be considered as the outlines of an animal which the Seminal Liquor is still sketching', and such a claim would in Reid's eyes have smacked of materialism.[17]

Buffon's conception of organic matter contradicted Reid's own position in three respects. First, Buffon seemed to hold that the power of organization was intrinsic to matter. While he drew an analogy between this organizational power and gravity, Buffon did not say that it resulted from the operation of an impressed immaterial principle as would an orthodox Newtonian, and hence he could be read as attributing active powers to matter itself. Reid's fellow Scot, the Edinburgh Professor of Natural Philosophy, John Stewart, interpreted Buffon this way, for he wrote in 1754 that Buffon and his collaborator John Turberville Needham had 'carried the activity of matter to the highest pitch'. Reid always distinguished sharply between inert matter and the forces impressed upon it, and it is likely that he too believed that Buffon had rejected this distinction.[18]

Secondly, Buffon assumed that matter was capable of self-organization, as had earlier materialists such as Anthony Collins.[19] Reid emphatically rejected this view. He insisted that even the most rudimentary form of organization was the result of intelligence, rather than the blind operations of material causes, and he stressed that this was a point of 'common sense'. That is, Reid maintained that it is a first principle of our reasoning that 'design, and intelligence in the cause, may be inferred, with certainty, from marks or signs of it in the effect'.[20] Because plants and animals, even in embryonic form, bear the marks of design, Reid thought that it followed *necessarily* that they are the effects of an intelligent cause, and hence could not be produced by matter and its attendant powers alone. In so far as Buffon's theory of generation presupposed that matter could organize itself, it contradicted one of Reid's basic principles of common sense and, what was more, it countenanced materialism.

A third feature of Buffon's theory of generation which Reid found disturbing was the French savant's dismissal of final causes. In the *Histoire naturelle*, Buffon critically reviewed the various strategies available for explaining reproduction, and in so doing he renounced any appeal to divine design.[21] To a pious philosopher like Reid, this verged on atheism. He later warned his Glasgow students that 'Since the time of des Cartes . . . we find that some have adopted his sentiments [concerning final causes] who may be suspected of a tendency to Atheism of These we may reckon Maupertuis and Buffon'.[22] Moreover, as I have already mentioned,

Buffon denied that generation was in any way dependent on divine power or activity, and he declared that hypotheses which invoked God's action simply precluded any genuine causal inquiry.[23] Buffon's theory of generation thus conjured up a vision of the natural order stripped of divine design and purpose. Nature, he seemed to say, was self-sustaining and self-regulating because matter was perpetually active and continually organizing itself into animate beings. Consequently, Reid had more than enough reason to suspect Buffon of materialism as well as atheism.[24]

Reid seems to have found Buffon's analysis of the mental operations of man and the higher animals suspect as well. In the *Histoire*, Buffon pictured animals as being mere automata, lacking the faculties of memory and reason, and acting according to mechanical principles.[25] Such a characterization was anathema to Reid because he believed that it ultimately led to materialism, and his reading notes transcribed below suggest that he was troubled by Buffon's mechanistic conception of the behaviour and mental functions of animals.[26] Reid was worried too by Buffon's depiction of humankind, and in the light of later remarks one entry is especially significant: 'The Sense most relative to knowledge is Touch which man has in a Superior degree to all animals'.[27] At some point in the late 1750s or early 1760s Reid detected in this claim of Buffon's a hint of that brand of materialism expounded in Helvétius's notorious *De l'esprit*, published in 1758. Having so decoded Buffon's conception of humankind, Reid attacked both Helvétius and Buffon in his Glasgow lectures. He warned his students of the

> extravagant Opinion advanced by Helvetius sur l'Espirit and countenanced by Buffon in some parts of his natural Historie That the chief Difference between Men and Brutes lies in this that Nature has given to Men finer Organs of Touch particularly in the hands and fingers. That Men having by this means a more accurate knowledge of the tangible qualities of Bodies, all their other improvements are owing to this. According to this System it might be said that a mans Wit lies in his finger ends.[28]

However, Reid did not, in the end, think that Buffon was a thoroughgoing materialist like Helvétius. Citing Buffon's account of the dissection of the brain of an orang-utan, Reid told his Glasgow students (with a touch of irony) that on this occasion [Buffon was] struck with a fit of Religion and acknowledge[d] that the human Understanding did not arise from organization since this Animal so similar to Man in its Organization had not the least similitude to man in its Understanding.[29]

Reid thus acknowledged that Buffon allowed that there is an immaterial sentient principle in man, if not in any other species of animal. But Reid's comments here also indicate that he regarded Buffon's attendance at the dissection of the orang-utan as probably the *only* occasion when the great French naturalist had been 'struck with a fit of Religion'.

Taken in conjunction with his later manuscript and published writings, Reid's reading notes from the *Histoire naturelle* reveal that while he may have been prepared to draw on the *Histoire* for factual information, he was adamantly opposed to Buffon's critique of taxonomic systems, his characterization of mathematics and its role in the physical sciences, his cosmogony, and his theory of generation. Passages in Reid's *Inquiry*, his Glasgow lectures, and his *Essays on the intellectual powers of man* show that he suspected Buffon of atheism and materialism, and it would seem that Reid endeavoured to discredit Buffon's heterodox ideas by exposing his theories of the earth and of generation as being no more than a mélange of ill-grounded hypotheses and conjectures. In so far as Reid continued to read the *Histoire naturelle* after his initial confrontation with the first three volumes of the work in the 1750s, there is little doubt that the *Histoire* had a lasting impact on his own philosophical outlook.

While Reid's interests in natural history were (on the whole) varied and changeable, his unpublished papers reveal that for most of his career Reid was preoccupied with the phenomena of generation. Apart from a passing reference to the theory of *emboîtement* in the *Essays on the intellectual powers of man*,[30] however, Reid left no published record of his thoughts on the long-standing debate over the propagation of plants and animals. It was only when Lord Woodhouselee printed selections from the Reid–Kames correspondence in his *Memoirs of the life and writings of the Honourable Henry Home of Kames* (1807) that Reid's own speculations on this topic first appeared, for Woodhouselee included a letter to Kames, probably written in 1775, in which Reid detailed his theory of organized atoms. Woodhouselee compared this theory with that of the noted Swiss naturalist Charles Bonnet, but this comparison was later rejected by Sir William Hamilton, who claimed that 'Reid's opinion has comparatively little resemblance to the *involution* theory of Bonnet'. In line with his own scholarly interests, Hamilton preferred to stress the similarities between Reid's theory and the panspermatism of the ancient Greeks and 'the recent physiological speculatists of Germany', while at the same time excusing it as a 'curious . . . solitary escapade of our cautious philosopher in the region of [the] imagination'.[31] In fact, the comments of both men are highly misleading. Although there are parallels between the theories of Reid and Bonnet, Woodhouselee's suggestion that Reid derived his ideas from the works of Bonnet cannot be sustained in the light of the available evidence detailed below. Moreover, *pace* Hamilton, Reid's letter to Kames was more than a 'solitary escapade', for his manuscripts show that it was the product of some forty years' reflection on the question of generation.

The earliest reference to the subject of generation in Reid's papers occurs in the minutes of the 12 January 1736 meeting of the Philosophical Club, when Reid and his fellow members discussed the question 'What Things in the Course of Nature we may reasonably ascribe to the continual influence & Operation of God or other active powerful and Invisible beings under him?' The participants apparently agreed that in general 'The Animal Oeconomy requires a Constant Supply of force to carry on its Motions', and that generation as well as voluntary muscular motion were specific instances wherein some kind of active power or powers exercised 'A Continual Influence'.[32] Significantly, they also concluded that the 'Animacular Hypothesis [was] not probable', presumably because they preferred the rival ovist hypothesis which was more widely accepted at the turn of the eighteenth century. Given the natural theological context of their discussion, it may be that Reid and his colleagues objected to animalculism on the grounds that it was inconsistent with the wisdom and design exhibited by nature since it implied that there was an incredible wastage of rudimentary organisms in the processes of generation.[33]

After a gap of some sixteen years the subject of generation reappears in Reid's manuscripts, this time with reference to his teaching responsibilities at King's College. In his 'Scheme of a Course of Philosophy' drawn up in 1752, Reid made provision for lecturing on the generation of animals, and it would seem that he did in fact do so following the restructuring of the curriculum at King's in 1753. Although the texts of his lectures have not survived, we know that he defended the theory of pre-existence in these lectures and that he included materials taken from the works of Hooke, Leeuwenhoek, Réaumur, Swammerdam and the English microscopist Henry Baker.[34] In addition, Reid may have incorporated some of the findings of Buffon, despite his misgivings about the materialist overtones of Buffon's explanation of generation.

During the 1750s Reid confronted the problem of spontaneous or equivocal generation as well, for on 28 June 1758 the members of the Aberdeen Philosophical Society considered the question posed by George Campbell, 'Can the Generation of Worms in the Bodies of Animals be accounted for on the common Principles of Generation?'[35] From the end of the seventeenth century onwards, it was widely recognized that the generation of intestinal parasitic worms could not be reconciled easily with the principles of preformationism. By contrast, Buffon and his collaborator John Turberville Needham argued that the generation of such parasites could be fully explained by the theory which they advocated, and materialists cited the phenomena of equivocal generation to prove that matter was capable of reproducing itself without the supervention of immaterial causes.[36] Issues such as these probably inspired Campbell's question, and we can see from the records of the discussion in the papers of David Skene that Campbell and his colleagues believed that Buffon and Needham had attempted experimentally 'to revive the exploded Doctrine of Equivocal Generation or at least afford an instance of Animals being produc'd in such Circumstances, as the receiv'd Opinion of the Generation by Eggs, does not seem capable of admitting'.[37] The Aberdonians thus understood that the experiments of Buffon and Needham challenged accepted theory and gave credence to the suspect notion of equivocal generation, but the extant documents unfortunately do not indicate whether the members of the Society thought that these experiments were in fact explicable in accordance with 'receiv'd Opinion', and Reid's papers are likewise silent on this issue.

Although Reid was no longer obliged to teach the theory of generation after taking up the Glasgow Chair of Moral Philosophy in 1764, he nevertheless chose to consider the subject briefly in his classes on natural theology and on the culture of the human mind. From a set of student notes on natural theology taken in the 1766–7 session, we can see that Reid reviewed some of the general facts about the propagation of vegetables and animals, emphasized the wisdom and design illustrated by these facts, underlined the folly of materialism, and asserted that the seed of plants 'contain the future plant in Miniature'.[38] Similarly, in his private lectures on the culture of the human mind at this time, Reid quickly surveyed the rival systems of animalculism, ovism and epigenesis, and said that it was probable that the foetus existed prior to conception, without committing himself on whether it was contained in the male semen or the female egg.[39] Reid's scepticism regarding our capacity to explain the phenomena of generation in these lectures is striking, and it is difficult to avoid the conclusion that his sceptical posture was largely a defence against materialism. For by arguing that our knowledge was limited to a highly circumscribed range of 'facts', Reid denied the validity of materialist discourse about the efficient causes of reproduction. He thus tried to neutralize the position of Buffon and others by declaring that many of the problems they raised were beyond the compass of the human mind.[40]

Outside of the classroom, Reid occasionally engaged in the empirical investigation of some of the more puzzling phenomena related to generation. In 1767 he performed some rudimentary experiments suggested to him by David Skene, and the air of excitement in his letters to Skene indicates that they were of considerable theoretical interest.[41] Reid's comments show that he was trying to discover whether or not the animalcules he obtained from placing a tree fungus in purified water would be transformed into mushrooms, and thereby demonstrate that vegetable matter could be transformed into animal, and vice versa. The exact nature of Skene's and Reid's theoretical interest in the results of this experiment are unclear, but it should be noted that in the 1740s John Turberville Needham had adduced evidence to show that such transformations took place in the course of outlining his epigenetic account of generation, and that Lazzaro Spallanzani had attacked Needham's arguments in his *Saggio di osservazioni microscopiche* of 1765. It may well be, therefore, that Skene and Reid were keen to try 'Linnaeus's experiment' because of the light it could shed on questions being debated by the proponents of epigenesis and preformation.[42]

Encouraged by Lord Kames, Reid outlined his thoughts on the propagation of plants and animals in letters dating from the period prior to the publication of Kames's *The gentleman farmer* in 1776. Only a part of their correspondence dealing with this topic survives, but in the letter published by Woodhouselee and in a more detailed draft of it to be found in the Birkwood Collection, Reid stated the essence of his theory of generation.[43]

According to Reid, animals and vegetables 'are at first organized atoms' in which all their parts exist and are enfolded.[44] These atoms are diffused throughout nature and behave like ordinary matter, remaining dormant like seeds until they are 'brought into their proper matrix or womb', where 'they are commonly surrounded with some fluid . . . in which they unfold and stretch themselves out to a length and breadth perhaps some thousand times greater than they had when folded up in the atom'.[45] He reconciled this account of the unfurling and expansion of the atom with Harvey's observation that in animals an embryo cannot be seen in the fluid which eventually appears in the womb after conception, by arguing that the foetus is invisible because its parts 'are yet so immensely thin that they are quite pellucid', and hence indistinguishable from the fluid which contains them.[46] Moreover, Reid believed that the rapid appearance of the *punctum saliens*, brain, and spine in the fluid found in the womb could only be explained as being the result of the 'sudden unfolding' of the organized atom, since it was 'very improbable' that an embryo 'should by growth & nourishment increase its Size many millions of times in the space ⟨of⟩ two or three days before it is visible and after it becomes visible should never double its size in that short space'.[47]

In the draft of his letter to Kames, Reid also remarked on the propagation of the polyp and on the question of spontaneous generation. Materialists and critics of preformationism had fully exploited Abraham Trembley's discovery in 1740 of the self-propagating power of the polyp, whereas their opponents were intially thrown into confusion.[48] Explanations of the polyp's ability to regenerate consistent with the principle of pre-existence were, however, soon being offered, typically in terms of 'germs' scattered throughout the body of that organism. It is unclear from Reid's letter whether he adopted this explanatory strategy wholesale, but he did claim that 'as in most Vegetables every part of the [polyp] may serve as a Womb to the organized Atom'.[49] Regarding the series of experiments performed by Buffon and Needham in which 'animalcules' were produced in pepper water which had been boiled and placed in hermetically sealed flasks, Reid wrote that 'all I would infer from this would be that the organized Atoms of these Animals were in the pepper Water & that it was a proper womb for their expansion and growth'. To the objection that the boiling of the pepper water would have destroyed the organized atoms, he replied that 'it is possible that boyling may no more hurt these Atoms than it hurts the Elementary Atoms of Water itself'.[50] Reid thus struggled to accommodate the findings of Buffon and Needham to his theory of organized atoms, but it must be said that his explanation of these anomalous experiments is far from convincing.

Having sketched his conjecture, Reid countered the criticisms levelled previously by his correspondent. Kames had claimed in an earlier letter that 'we must allow that certain particles may be endowed with a power to form in Conjunction an organized body'.[51] Reid responded that Kames had wrongly assumed that if we can distinctly conceive something then it must be possible, and he pointed to instances in mathematics which disproved this.[52] Secondly, he insisted that it was an axiom of common sense that wisdom and design like that manifested in the vegetable and animal kingdoms had to have been effected by a designing cause. He maintained that it was impossible that a plant or animal could have been 'formed by unthinking Matter guided by certain Physical Laws of Motion'.[53] It would seem that Kames had further charged that on the hypothesis of organized atoms, generation was a miracle because the Divinity perpetually intervened. Following Samuel Clarke, Reid argued that 'nothing is a Miracle which falls out according to the established order of Nature even tho' it should be the immediate operation of the Deity'. He added that we have no reason to believe that 'the Almighty finished all his Work at a stroke and has continued ever since an Unactive Spectator', since we cannot prove that such a mode of divine governance was best or even practicable.[54]

Unmoved by these arguments, Kames attacked the theory of organized atoms along with those of *emboîtement* and pre-existence in *The gentleman farmer*. Kames claimed that all of these related explanations of the generation of plants were conceptually incoherent, and he quoted extensively from a letter he had received from the botanist the Revd John Walker, in which Walker rejected the notion of organized atoms and argued for the principle *omne vivum ex ovo*. Walker was, in fact, commenting on Reid's ideas when writing to Kames, and had this to say about Reid's hypothesis:

> As for Dr Reids Idea of organized Atoms, diffused at large through the Universe, & detached from all Animal & Vegetable Bodies, it is not countenanced by any thing within the Sphære of my Knowledge. He adduces no Facts, nor do I recollect any to support it. I would wish to know, his Illustration of it, as our worthy Friend, is not one who is ready to assume things upon slender Grounds.

When publishing this letter, Kames carefully edited out the references to Reid, and made it appear as if Walker was criticizing the theory of organized atoms without any particular proponent of that theory in mind.[55] It is unclear whether or not Reid knew of this subterfuge, but he cannot have derived much satisfaction from Kames's text in any case, because he undoubtedly felt that his patron's account of the propagation of plants was too much like that of Buffon, and that Kames had again left himself open to the charge of materialism. For Kames had claimed that the 'power[s] of gravity, of resistance, of continuing motion [are] *essential* to matter in general', and that 'the power of producing a body similar to itself, [is] *inherent* in all organic bodies'. Instead of preserving a sharp distinction between matter and powers such as gravity, Kames seemed to ascribe the powers of unorganized and organized bodies to matter itself, and even though he insisted that God had originally given these powers to matter, men schooled in the texts of Newton and Clarke like Reid saw Kames's position as being tantamount to materialism.[56]

At some point in the 1770s Reid began to put forward the hypothesis he communicated *circa* 1775 to Kames in his lectures on the culture of the human mind, for we find the following addition to the text dealing with the theory of generation discussed above: 'Conjecture that Animals & Vegetables originate from Organized Atoms'.[57] This insertion implies that Reid's views had changed since he had originally written this lecture, and this raises the question as to why the shift occurred. Was it, as Woodhouselee suggested, because Reid had read Bonnet?

The set of detailed reading notes from Bonnet's *Contemplation de la nature* (1769) and *Palingénésie philosophique* (1770) transcribed below (MS 2131/3/II/16) lends prima-facie support to this interpretation, but the notes are undated, and it is arguable that they were taken in the late 1770s or early 1780s. The order in which these notes were written shows that Reid initially read the *Contemplation*; the earliest he could have done so was June 1774, for it was then that he first seriously studied Hartley, to whom he compared Bonnet in his notes.[58] However, a later date seems more probable, since the manuscript begins with a cancelled passage entitled 'Of the Subjects of Active Power', which is most likely related to Reid's discourse on the question 'Have we any Reason to ascribe Active Power to Beings not endowed with Understanding and Will?', which he read before the Glasgow Literary Society in November 1777.[59] As for the *Palingénésie*, the bibliographical details of the copy which Reid used clearly indicate that it was from the duodecimo edition of Bonnet's *Oeuvres* which appeared between 1779 and 1788.[60] Moreover, Joseph Priestley quoted from the *Palingénésie* on the title page of his *Disquisitions relating to matter and spirit* (1777), and it may well be that his citation finally prompted Reid to read Bonnet with some care.[61]

Questions of dating aside, Lord Woodhouselee was no doubt unaware of Reid's long-standing interest in generation, and assumed that the ideas expressed in the letter to Kames must have been derived from elsewhere. Since Kames cited Bonnet in *The gentleman farmer* as a proponent of the system of organized atoms, it was reasonable for Woodhouselee to infer that Bonnet was Reid's source.[62] Yet once the background to Reid's letter is reconstructed, and his continued commitment to the theory of pre-existence is revealed, Woodhouselee's reasoning becomes less persuasive.

Moreover, the conceptual differences between the hypotheses of Reid and of Bonnet imply that Reid developed his own independently. Bonnet, for example, explained the regeneration of the polyp in terms of germs diffused throughout its body, whereas we have seen that Reid was vague on this point and argued instead that each part of the polyp could act as a matrix for an organized atom. Bonnet too was a leading advocate of the theory of *emboîtement*, while Reid avoided speculation on the origins of organized atoms. Furthermore, Reid's reading notes from the *Contemplation de la nature* (to which Woodhouselee did not have access) suggest that he found certain aspects of Bonnet's system highly suspect. While he clearly approved of Bonnet's piety and the Swiss naturalist's critique of Buffon's notion of 'organised Molecules', he evidently thought that Bonnet's theory of human nature verged on materialism, and he observed that Bonnet (like Hartley before him) had been led into fatalism because he had posited a causal connection between motions in the nerves and mental processes.

Reid probably also had reservations about Bonnet's conception of matter, for he noted that Bonnet believed that the basic units of matter are active and that all created beings have some degree of sense and perception. Reid would not have endorsed either claim, since he insisted that all matter is inert and that unorganized bodies are senseless. He may well have been critical too of Bonnet's use of the terms 'irritation' and 'stimulus' when explaining the generation of plants and animals, for in one of his late Glasgow Literary Society discourses on materialism, he attacked those 'modern Physiologists' who employed these terms to refer to efficient causes (as he apparently judged Bonnet to have done), and argued that they properly denoted effects whose cause lay outwith the bounds of human knowledge. Finally, although Reid thought that Bonnet had 'very just Notions of the Instinct of Animals', he can hardly have approved of Bonnet's almost Malebranchian 'Conjecture' that some insects and animals have special fibres 'which at the proper Season Suggest certain Ideas to their minds of the works they are to produce and the means of producing them'. It is unlikely that Reid would accept that the fibres of the brain, the nerves, or the muscles could actually suggest ideas to the mind, and in any case he would have regarded such speculations as idle, if not pernicious, since instincts were for him God-given principles of action and belief whose underlying mechanisms were unknowable. The available evidence thus suggests that Reid should not be seen as a Scottish disciple of Bonnet. He seems to have evolved his theory of organized atoms prior to his reading of Bonnet and, as I have pointed out, there are significant conceptual differences between their two systems. More importantly, even though Reid found the religious tenor of Bonnet's writings appealing, it is arguable that he discovered that there was some justice in the charges of materialism that Bonnet had been so anxious to rebut. However, Reid's papers unfortunately provide no clue as to precisely when and why he modified his earlier views on the theory of generation.[63]

When discoursing on the inadequacies of materialism before the Glasgow Literary Society in the 1780s or early 1790s, Reid touched briefly on the subject of generation, and here again he stressed that the 'ways by which animals and vegetables produce their kind are various and all equally mysterious & incomprehensible to human understanding'.[64] As I have indicated, Reid's assertion that we are ignorant of the processes of reproduction can be interpreted as an epistemological strategy designed to rule out the possibility of materialist explanations of generation, and it is therefore not surprising to see Reid resorting to this tactic in the context of his response to Joseph Priestley. Priestley's materialism also prompted Reid to analyse the functions of what he called the 'principle of life' in the vegetable and animal kingdoms. Reid had earlier mentioned this principle in passing in his correspondence with Lord Kames, but he now detailed its role in carrying out the vital functions of plants and animals, and argued that the phenomena of generation demonstrated the existence of an immaterial power which enabled organic beings to reproduce after their kind. Reid recognized that what he had to say about the unity of the principle of life seemed to contradict what was known about the reproductive capacities of some animals and vegetables such as the various kinds of polyps, yet he was adamant that the facts had to be reconciled with the 'principle of reason' that the vital principle was indivisible. In order to do this, Reid suggested that each and every part of these plants and animals contains a 'living Foetus' possessing its own principle of life and capable of growing to maturity if separated from the parent. What is notable about Reid's discussion of the phenomena of regeneration is that for metaphysical reasons he was determined to preserve the unity of the principle of life in the face of anomalous empirical evidence; furthermore, his explanation of the phenomena differed significantly from that which he had advanced in his letter to Kames and was much closer to that given in the works of Bonnet.[65]

Reid's papers on natural history well illustrate both his strengths and his limitations as a naturalist. To his credit, he had an insatiable appetite for the study of God's creation, which was fuelled by his religious sensibilities, his desire for economic improvement and his boundless curiosity about the world around him. He brought to his natural historical studies an acute mind, and his response to Buffon's *Histoire naturelle* shows him capable of grappling with a wide range of empirical and metaphysical issues. Yet it must also be said that his work on generation reveals that Reid was all too willing to withdraw into a posture of nescience about the processes involved in reproduction in order to protect his religious beliefs, and his anxieties about materialism led him to ignore questions which he should perhaps have addressed. Thus Reid never seems to have worked out his theory of organized atoms in any great detail, and in his writings which have survived he simply evaded the empirical and conceptual problems facing his hypothesis by declaring that the origins of living things lay beyond the bounds of human understanding. Such protestations were certainly an effective way of foreclosing debate, but it is not difficult to sympathize with Lord Kames, who charged that Reid's methodological attitudes tended to '[damp] the spirit of inquiry'.[66]

Reid's papers also help to substantiate the point made by Paul Lawrence Farber that in the eighteenth century the field of natural history was made up of 'divergent traditions sharing a general subject matter (the living world) and a set of catholic (with a small «c») naturalists'. According to Farber, there were four such constitutive research traditions in this period, devoted to nomenclature and taxonomy, descriptive natural histories, the elucidation of animal form through comparative anatomy and the experimental and comparative investigation of plant and animal physiology. Although Farber claims that individual naturalists tended to work in one of these areas to the exclusion of the others, in Reid's case we can see that his teaching, reading and research were informed by all four of these traditions.[67] Thus, in the course of his lectures Reid adopted the roles of the taxonomist, the descriptive historian, the comparative anatomist and the physiologist, and there is no sign that he found these roles to be incompatible. Reid was equally familiar with the writings of classifiers like Linnaeus and those of Buffon and, although he seems not to have read widely on comparative anatomy, he was reasonably well acquainted with works on physiology. Influenced by his close friend David Skene, he was deeply interested in the problems involved in systematic taxonomy, although he also recognized the value of careful descriptive histories and, like Skene, he attempted to extend the boundaries of classification and description to encompass the study of the human mind.[68] Yet we have seen that Reid's own researches were not confined solely to the classification of the creation, for he was a serious student of plant and animal physiology, and his efforts to understand the problematic phenomena of generation represent one of the more important chapters in his intellectual biography. Thus Reid's manuscripts encapsulate the complex internal structure of eighteenth-century natural history, and they serve to remind us of the broad range of metaphysical, theoretical and empirical issues addressed by natural historians in the Enlightenment.

2. Physiology

Muscular motion was a popular, not to say fashionable, subject of speculation in the eighteenth century, its vogue encouraged to some extent in Britain by the annual lectures on this and related topics instituted at the Royal College of Physicians and the Royal Society through the bequest of the Restoration physician William Croone.[69] It is not surprising therefore to find that Thomas Reid devoted some thought to this aspect of the animal oeconomy. It was, however, more than simply an intellectual diversion for him, since the phenomena of muscular motion raised crucial issues concerning the powers operative in the animal creation, and the relationship between mind and matter.

Reid's 'Minutes of a Philosophical Club' from 1736 provides the earliest hint of his interest in the problem of muscular motion, for part of the entry from 12 January reads: 'The Animal Oeconomy requires a Constant Supply of force to carry on its Motions[.] Other instances of A Continual Influence [of God or other spiritual agents governed by Him] – in Voluntary motion'. Unfortunately there is no indication as to whether Reid and his colleagues in the club agreed with their fellow Aberdonian Andrew Baxter that all animal motion was divinely caused, or whether they ascribed vital and voluntary motions to the operation of 'other active powerful and Invisible beings'.[70] What *is* clear from this passage is the natural theological use to which the phenomena of muscular motion were put. Starting from the premise that matter is essentially inert, Reid and his associates were able to infer that there must be some active, immaterial cause of animal motion, and hence controvert the claims of proponents of materialism. Thus while physiologists who assumed the passivity of matter struggled to find a convincing naturalistic explanation of muscular motion, Christian apologists could exploit the phenomena for their own ends.[71]

Reid returned to the subject of muscular motion in his teaching at King's College Aberdeen. Following the curriculum reforms of 1753, he apparently included in his natural history course three lectures based on Robert Whytt's *Essay on the vital and other involuntary motions in animals* (1751), which were intended to demonstrate that 'The Origin of Motion in the human Body . . . is not Mechanical'.[72] Whytt's *Essay* well suited this aim, for he there argued that

> the human body ought not to be regarded (as it has too long been by many physiologists) as a mechanical machine, so exquisitely formed, as, by the mere force of its construction, to be able to perform, and continue the several vital motions; actions far above the powers of mechanism! But as a system, framed indeed with the greatest art and contrivance . . . yet a system whose functions are all owing to the power and agency of an immaterial sentient principle to which it is united, and by which every part of it is animated and put in motion.[73]

Whytt attacked iatro-mechanists, Cartesians and materialists alike for conceiving of the human body as a mere machine, and it is likely that it was his intense anti-mechanism and anti-materialism which recommended Whytt's physiology to Reid.[74] In addition, Reid would no doubt have approved of the Newtonian conceptual framework of the *Essay*, and Whytt's emulation of Newton's cautious inductivism, anti-hypotheticalism and avoidance of undue speculation about efficient causes.[75] Finally, Reid would have whole-heartedly endorsed Whytt's conclusion that 'true physiology' both proves the existence of an immaterial soul, and 'leads us up to the first cause and Supreme Author of all'.[76] It was probably for reasons such as these that Reid declared Whytt's theory of muscular irritability 'the most probable Hypothesis yet advanced on [the involuntary motions of animals] & I think the onely one that deserves examination'.[77]

Yet Reid was not entirely uncritical. Although he recognized that Whytt's theory was confirmed by numerous instances of muscular motion being induced by various stimuli, he wrote that 'it must be acknowledged that there are considerable difficulties attending this Hypothesis'. In particular, he thought that the word 'stimulus' implied conscious perception. He observed that 'we can affix no other Idea to the word Stimulus but that of Pricking or the Sensation arising from pricking. We do not commonly call it a Stimulus where both [of] these do not go together'. Consequently, it seemed to him to be an abuse of language to say that the circulation of the blood or the motion of the lungs is caused by a stimulus since we are not aware of any stimulation in either case. Moreover, the notion of 'a Stimulus that is not felt or perceived' seemed to him to 'border upon a Contradiction'. Reid proceeded to propose three alternative solutions to this conceptual puzzle. First, he allowed that the term 'stimulus' might be used to 'signifie onely an unknown somewhat like an occult Quality', by which he perhaps meant that 'stimulus', like 'attraction' or 'gravitation', should be regarded as a theoretical term referring solely to observable effects rather than to efficient, physical causes. Secondly, he suggested along Stahlian lines that 'there is a Sentient Principle in us distinct from the Mind which feels this Stimulus and is affected by it'. This solution, he believed, had the advantage of avoiding the linguistic impropriety involved in implying that 'mere inert Matter' can be 'acted upon by a Stimulus'. Thirdly, he considered the possibility that 'tho' the Mind ha[s] in the beginning a perception of this Stimulus yet like other feelings of pleasure and pain it becomes insensible by habit'. He rejected this outright on the ground that 'where a Stimulus by habit & use becomes insensible it looses its effect'.[78] Because his discussion breaks off at this point, it is impossible to determine whether Reid actually managed to reason his way out of this conceptual impasse. However, we shall see that he did not in any case entirely abandon the concepts of stimulus and irritability, although he remained vigilant in guarding against what he perceived as abuses of these terms.

As part of his study of the human senses and his critique of the theory of ideas, Reid was also interested in the physiology of perception during his years at King's College, if not before. The significance of his work on this subject is fourfold. First, it should be emphasized that Reid believed that the theory of ideas was embedded in a physiological account of human perception, since it presupposed an explanation of how images of external objects were transmitted to the brain and resulted in ideas perceived by the mind.[79] Although his interpretation of the theory of ideas has recently been contested, we can begin to see why Reid understood the theory in the manner that he did if we look at the intermingling of physiological and philosophical considerations in the works of Descartes, Malebranche and Locke, whom Reid regarded as the founders of the modern science of the mind.[80]

Descartes' theory of the mind, for example, was premised on a fairly detailed physiological analysis of the functions of the human nervous system and the role played by animal spirits in sensory perception, memory, imagination, the arousal of the passions, and the general temperament of the individual.[81] Moreover, in *Traité de l'homme* Descartes went so far as to suggest that in perception, the motions of animal spirits leave patterns on the surface of the pineal gland, and that these patterns constitute 'figures' or 'ideas' which the soul 'consider[s] directly'.[82] Indeed, it was because Descartes conceived of ideas in this way that he located the seat of the soul in the pineal gland, since that was the only place in the brain where the two images from the eyes or the twin impressions on the nerves from the other senses could be united and exhibited to the soul as one.[83] Clearly, Descartes' reasoning only makes sense if we suppose that ideas have some physical basis, and it was precisely this combination of physiology with what we now regard as 'philosophy' that shaped Reid's understanding of the theory of ideas.[84]

The writings of Nicholas Malebranche display a similar mix. In *De la recherche de la vérité*, we find a blend of physiological speculation and philosophical analysis, as Malebranche relied on the theory of animal spirits to explain muscular motion, perception, the operations of the imagination, individual and national characters, habit and the passions.[85] Even though Malebranche denied that physiological processes were the efficient causes of mental events, he still insisted that the former occasioned the latter and that their correlations were governed by 'the general laws of the union of soul and body'.[86] Consequently, physiology still played a crucial role in Malebranche's theory of the mind, even though his conception of the relations between mind and body was radically different from that of Descartes.

Turning to John Locke's *Essay concerning human understanding* (1690), the overt connections between physiology and philosophy were less close than in the writings of Descartes and Malebranche, in so far as Locke disclaimed any interest in 'the Physical Consideration of the Mind', such as 'by what Motions of our Spirits, or Alterations of our Bodies, we come to have any Sensation by our Organs, or any *Ideas* in our Understandings; and whether those *Ideas* do in their Formation, any, or all of them, depend on Matter, or no'.[87] Yet despite Locke's formal disavowal of physiological theorizing, there are a number of passages in the *Essay* which hint at the physiological mechanisms upon which his account of perception and the other faculties of the mind rested. Locke claimed, for instance, that our sensations are 'produced in us, only by different degrees and modes of Motion in our animal Spirits, variously agitated by external Objects', and he invoked Malebranche's physiological account of habit to explain the association of our ideas. Locke too drew on the ideas of Descartes and Malebranche when he suggested that memory depends on there being physical traces in the brain, and in a striking image he compared the ideas stored in the memory to inscriptions on tombstones which deteriorate through time.[88] For those predisposed to find them, the rudiments of a physiology of mind were thus to be found scattered throughout the *Essay*, and it is highly significant that Reid implied that Locke's account of perception rested on a physiology virtually identical to that of Descartes.[89] It would seem, therefore, that Reid believed that Locke had combined physiology with philosophical analysis in essentially the same way that Descartes and Malebranche had done, and hence that the theory of ideas, as formulated by this triumvirate, was rooted in a set of physiological hypotheses regarding the mechanisms involved in human perception. Consequently, Reid's attack on the theory of ideas was aimed as much at the physiological presuppositions of that theory as it was at the philosophical errors which he thought it perpetrated.[90]

Secondly, Reid's critique of hypotheses in the *Inquiry* was addressed in part to those physiological theories of perception which assumed the existence of animal spirits or elastic ethers, or the capacity of the nerves to vibrate like the strings of musical instruments. After moving to Glasgow in 1764, he continued to dismiss these theories as mere hypotheses in his lectures, and we shall see that some time after 1774 he extended his criticisms to the etherial physiology of David Hartley.[91] Reid's anti-hypotheticalism, therefore, evolved in the context of his review of the physiological mechanisms used to explain the physical basis of human perception.

Thirdly, Reid's criticisms of these mechanisms are symptomatic of the general reaction (discussed by Schofield and Brown)[92] against the mechanistic and reductionist mode of physiological theorizing prevalent in the first half of the eighteenth century. Robert Whytt was one of the first physiologists in Britain to reject ethers and animal spirits, and to question whether the phenomena of the human body were explicable in terms of the attractive and repulsive forces active in the inanimate realm of nature. Reid likewise assailed the standard explanatory concepts of the physiologists' repertoire, and asserted that 'in the vegetable and animal kingdoms, there are strong indications of powers of a different nature from all the powers of unorganized bodies'.[93] Thus Reid's assault on nervous ethers and fluids and his opposition to reductionism reflect broader theoretical and methodological shifts within the science of physiology, as well as trends in eighteenth-century natural philosophy more generally.

Fourthly, it was through his work on the physiology of perception that Reid encountered the ideas of one of the leading exponents of animism in mid-eighteenth-century Scotland, William Porterfield. In his 'An essay concerning the motions of our eyes' and later in his *Treatise on the eye*, Porterfield argued that all of the vital motions of the human body are originally voluntary and consciously controlled by the mind, and that it is only through custom and habit that these motions ultimately become unconscious, though still mind-governed. Porterfield's views were subsequently attacked by Robert Whytt, who referred to his Edinburgh colleague as a 'subtle defender of . . . Stahlian doctrine'.[94] It is likely that Reid shared Whytt's reservations about Porterfield's account of the vital motions, given his sympathy for Whytt's position and his critical attitude towards Stahlian physiology. Moreover, while Reid greatly respected Porterfield's writings on physiological optics, he marshalled together a variety of visual phenomena in the *Inquiry* in order to refute Porterfield's explanation of single and double vision. Thus Reid had little truck with Porterfield's theoretical principles, even though he held 'the ingenious Dr Porterfield' in high esteem.[95]

Reid returned to the problem of muscular motion in his Glasgow lectures on natural theology and on the fine arts. In the former, muscular motion was one of a whole range of phenomena Reid used to illustrate our nescience about the inner workings of nature. While he stressed that the structure of the nervous system and of the muscles manifested providential design, he was equally emphatic that the processes by which muscular motion is produced are above our comprehension. All that was known with any certainty, according to Reid, was that a power is transmitted from the brain through the nerves to the muscles which results in movement, and that this power is not conveyed by a nervous fluid, as dissections had revealed that the nerves are solid. Beyond these facts, he affirmed, muscular motion remained a mystery. Here too, then, Reid arguably adopted a stance of nescience in order to combat those mechanistic and materialistic accounts of muscular motion which he opposed.[96]

Although Reid's prelections on the fine arts may seem an unlikely setting for a discussion of physiology, he in fact prefaced these lectures with a survey of the various relations of mind and body which included an analysis of voluntary and involuntary motion.[97] Man's voluntary motions, he told his students, posed the following question: 'How can acts of the mind produce any effect upon the body?'[98] He then presented two conflicting answers which had been given. The first was that the mind was not in reality the efficient cause of voluntary motions. If it were, some philosophers had argued, then we would be sensible of causing such motions, but since we do not perceive this, we must conclude that there is some other efficient cause, like the contraction of the muscles.[99] Secondly, Stahl had countered this position by asserting that all of our voluntary movements were indeed caused by the mind. He explained our unawareness of causing our movements in terms of the mind becoming habituated to, and hence becoming unaware of, its activity. Reid rejected both answers. Stahl's arguments were, he said, 'very ingenious', but habit was as difficult to explain as the phenomena Stahl used it to account for.[100] More importantly, Reid claimed that we must content ourselves with the fact, attested to by experience and common sense, that we can move our body by exercising our will. It was, he believed, 'a matter of no consequence but to gratify our curiosity how [voluntary motion] is performed'.[101]

Regarding involuntary motion, Reid again presented two opposing viewpoints which had been articulated by philosophers and physicians. On the one hand there were those like Stahl who held that all of the involuntary motions of the body are initially voluntary (and hence controlled by the mind), and that as these motions become customary we lose our consciousness of them. On the other hand, Descartes, Boerhaave and their followers had conceived of the body as a machine, and consequently had explained all involuntary motion in mechanical terms. According to Reid, this mechanistic hypothesis had 'been very fully refuted by Stahl and Dr. Whyte [i.e. Whytt] of Edinburgh', a statement which suggests that he was as anxious to discredit mechanistic conceptions of man in his Glasgow lectures as he had been in those delivered at King's College Aberdeen. However, Reid did not advocate any particular theory himself, since he thought that involuntary and vital motions 'arise entirely from the union of body and mind and are conducted in a manner of which we are totally ignorant'.[102] Within the context of his teaching at Glasgow, then, Reid attacked mechanistic physiology and criticized Stahlian animism, while maintaining a position of studied agnosticism concerning the causes of muscular motion.

Although Reid did not champion Whytt's ideas in his lectures, his discourse on the role of the 'principle of life' in animals and vegetables given to the Glasgow Literary Society in the 1780s or early 1790s shows that he continued to accept the essentials of Whytt's physiology. Reid here defined what he took to be the legitimate meaning of 'stimulus' and 'irritation', namely that these terms, like 'gravitation' and 'attraction', referred to effects manifest to the senses rather than to unobserved efficient causes. He was adamant that the notions of stimulus and irritation simply could not account for the various vital functions of animals and vegetables they were brought to explain, and that these functions were only explicable in terms of 'the Intercourse & Efficiency of some immaterial Cause'.[103] In this section of his discourse Reid may well have been implicitly attacking Whytt's erstwhile opponent Albrecht von Haller, who had postulated that irritability is a power possessed by the muscle fibres, or he may have been aiming his remarks at materialists like La Mettrie, who claimed that irritability was an innate property of matter.[104] Given his remarks on Bonnet's *Contemplation de la nature* discussed above, it may be that Reid also had the Swiss naturalist in mind when laying down his strictures on the use of these terms. His own view, as outlined in this discourse, was essentially that of Whytt, in so far as he stressed that the power involved in involuntary motions did not operate mechanically, and hence that this power was necessarily immaterial. His remarks in his lectures notwithstanding, Reid thus apparently still accepted the basic principles of Whytt's physiology, even though his endorsement of Wytt's system was perhaps less whole-hearted than it had been in the 1750s.

The culmination of Reid's speculations was his discourse read before the Glasgow Literary Society in 1795, 'Of Muscular Motion in the human Body', which was prompted by his 'Experience of some of [the] Effects of old age on the Muscular Motions'.[105] Developing a theme of his natural theology lectures, Reid devoted most of his discourse to showing how the musculature of the body 'is contrived with perfect Skill & Wisdom for the End for which it was intended'.[106] In addition, his anatomical excursus gave him the opportunity to illustrate the role of active powers in human physiology (and thereby further controvert Priestley's materialism). He suggested, for example, that the perfect 'Lubricity' of the tendons and their adjoining parts was the work of the 'principle of life', and he argued that the agency of immaterial powers in the human frame was further demonstrated by muscular motions which were either instinctive or habitual.[107] More importantly, he reflected at some length on the action of 'the Nervous Power or nervous Influence' conveyed by the nerves from the brain to the muscles.[108] As in his Glasgow lectures, Reid stated that we are largely ignorant about the manner in which voluntary motion is effected, and he reiterated his view that the laws of mechanics are completely insufficient to explain the operation of this power. Moreover, he again dismissed theories of animal spirits, vibrations of the nerve fibres and vibrations of an elastic physiological ether as mere hypotheses; they assumed entities which were not only incapable of causing the effects attributed to them, but whose very existence was unproven.[109]

Since the cause and *modus operandi* of the nervous power lay outwith the bounds of human knowledge, Reid declared that it was 'the business of the Philosopher' to observe 'every Circumstance relating to [this] unknown [power], that may enable him to judge of the effects that [it] may be expected to produce in certain circumstances, or may enable him to apply [it] to usefull purposes in human Life'.[110] In this spirit, Reid proceeded to describe those properties of the nervous power known to him. The first (and only one which need concern us) was that in voluntary motions this power acts only when excited by the exercise of our will. It was thus quite different from virtually all of the other natural powers studied by philosophers, which act constantly and are independent of the will of man and animal alike. The sole exception to this was the shock of the electric eel; Reid was evidently acquainted with the researches carried out in the 1770s by Sir John Pringle, John Hunter and Henry Cavendish, among others, on the electric eel or 'torpedo', and he too compared the eel's capacity to shock its prey with the exertion of the nervous power in voluntary motion.[111]

As for the connection between the will and the nervous power, Reid intimated that this was a question which could not be settled definitively. While he admitted that it was difficult to explain the nature of this connection, he claimed that we have a 'natural Conviction' that we are the cause of our voluntary actions, and he stressed that this conviction, 'which is the Work of Nature, and of the greatest Importance in Life, ought not to yield to Physical or Metaphysical Speculations'. In particular, he argued that even though Malebranche and the occasionalists had denied that the will is the efficient cause of our voluntary motions, we are nevertheless responsible for our actions, on the ground that 'he that believes a certain Effect to be in his Power and exerts his Power to effect it, is undoubtedly in moral Estimation the Cause of that Effect and accountable for it'.[112]

We see, therefore, that theories of voluntary muscular motion had definite moral implications, in so far as they had to provide some explanation of the relations between mind and body. Moreover, it is clear that Reid was himself attempting to steer a safe course between the Scylla of occasionalism and the Charybdis of mechanism. Occasionalism had been interpreted by its critics to entail that we are not morally responsible for our actions, and mechanistic models of man had been similarly perceived. By contrast, Reid championed a conception of man as a free and responsible moral agent, and hence his agnosticism about the precise connections between mind and body is perhaps best understood as part of his defence of what he took to be the common-sense view of man's moral agency. For by ruling out any theorizing about the relations between mind and body, Reid endeavoured to check both occasionalism and materialistic necessitarianism and thereby defend his belief in man's moral responsiblity.[113]

Reid's speculations on human physiology were thus conditioned by moral considerations throughout his career. From the 1730s onwards, he used the phenomena of muscular motion to demonstrate the existence of immaterial powers in the animal kingdom whose activity transcended the laws of mechanics. In the 1750s, Whytt's explanation of involuntary and vital motions provided Reid with a powerful weapon in his fight against mechanism and materialism, and although his attitude to Whytt's theory appears to have become increasingly ambiguous he seems to have continued to accept at least its fundamentals. Morally speaking, voluntary motion was perhaps the most sensitive topic, and here Reid sought a conceptual *via media* between the extremes of occasionalism and mechanistic necessitarianism. He emphasized our nescience concerning the physical connections between mind and body, in order to protect his belief in man's free will. As for the physiology of perception, it played a less prominent but still important role in Reid's intellectual development, since it was partly in the context of his critique of the theories then current that he formulated his own distinctive interpretation of Newton's First Rule of Philosophizing. These two strands of anti-materialism and anti-hypotheticalism converged in his response to Joseph Priestley, and it is to this subject that we shall now turn.

3. Materialism

Although Reid was concerned with a broad range of scientific, philosophical and mathematical problems during the middle decades of the eighteenth century, there is little doubt that in this period he was most exercised by the writings (as opposed to the person) of David Hume. In the 1770s, however, Reid's critical sights shifted south of the border as he responded to the challenge posed by the metaphysical and theological works of Joseph Priestley. Priestley's virulent attack on his epistemological principles deeply offended Reid, and Priestley's materialism and necessitarianism struck at the very heart of Reid's conception of the moral and natural order. Moreover, Priestley's appropriation of Newton in the materialist cause was for Reid tantamount to intellectual sacrilege, since he firmly believed that Newton had demonstrated the existence of a providential God and of immaterial agents active in nature, and that neither Newton's metaphysics nor his method gave any credence to heterodoxy in general and materialism in particular. Consequently, Reid's response to Priestley's materialism was as much an attempt to define what constituted the 'true' Newtonian system as it was a refutation of his opponent's matter theory, and his manuscripts thus provide a valuable insight into the conflicting images of Newton in eighteenth-century British natural philosophy.[114]

Reid's manuscripts also allow us to reconstruct a highly significant but little-studied episode in his career which forms a hitherto unrecounted chapter in the history of materialism in eighteenth-century Britain. In their respective biographies of Reid, Dugald Stewart and A. Campbell Fraser both mention Reid's engagement with Priestley's materialism in the 1770s and 1780s, but their analyses of Reid's response to Priestley's writings tell us more about their own philosophical priorities than they do about the actual course of events.[115] For all his merits as an annalist of the evolution of 'the Scottish philosophy', James McCosh similarly failed to cast any real historical light on Reid's reaction to Priestley's materialism. McCosh remarked on the care with which Reid prepared his manuscript 'Some Observations on the Modern [i.e. Priestley's] System of Materialism', and McCosh judged the 'Observations' to be 'of a thorough and searching character, distinguished for acuteness beyond almost any of the published writings of Reid, and written with great point and *naiveté*'.[116] Yet McCosh merely summarized the contents of the manuscript very briefly, and said nothing about its place in Reid's intellectual development at Glasgow. More recently, scholars have tended to focus on the epistemological issues at stake between Reid and Priestley, and their disagreement over free will, while virtually ignoring their differences over materialism.[117] It is thus all the more regrettable that in his admirable study of eighteenth-century British materialism, John Yolton ends his narrative with Priestley, and does not take into account Reid's attack on Priestley's reformulation of the materialist system.[118] Consequently, the manuscripts on Priestley's materialism and related topics transcribed below constitute an important new source for the history of the opposition to materialism in the eighteenth century and document an aspect of Reid's later career which has not received the scholarly attention it deserves.

Prior to the 1770s Reid was, as we have seen, preoccupied with the spectre of materialism in his physiological and natural historical inquiries, and various manuscript sources reveal that he was familiar with a number of key eighteenth-century texts in the materialist canon, such as Anthony Collins's exchanges with Samuel Clarke, Claude Adrien Helvétius's *De l'esprit* and probably Lord Bolingbroke's *Letters or essays addressed to Alexander Pope* (1754), as well as with the anti-materialist polemics of Clarke and his fellow Newtonian Humphrey Ditton.[119] The lesson he learned from Clarke and Ditton was that Newtonian natural philosophy was essentially antithetical to materialism, and Reid's reading of Collins, Helvétius and Bolingbroke no doubt underlined the fact that, in the first half of the eighteenth century, materialism was invariably bound up with Deism, atheism and aggressive anti-clericalism. Hence Priestley's attempted synthesis of materialism with the methodological principles of Newton and a highly rational form of Christianity was incompatible with Reid's understanding of the historical and logical relations between religious orthodoxy, Newtonianism and the materialist system.

Although Reid and Priestley were acquainted with each other's works by the late 1760s, they did not come into conflict until 1774, when Priestley announced in the Introduction to the third part of his *Institutes of natural and revealed religion* that he planned to publish a critical review of what he called the 'ill founded and dangerous' principles advanced by Reid and his compatriots James Oswald and James Beattie.[120] On 28 April 1774, Priestley sent advance copies of this Introduction to each of the Scottish savants, accompanied by a letter giving personal notice of his intentions. Oswald and Beattie courteously acknowledged Priestley's communication, but Reid refrained from replying.[121] Prompted by the letter, however, he looked through the first two volumes of Priestley's *Institutes* and, as his reading notes transcribed below show, Reid was decidedly unimpressed by his critic's performance. Passages on the capacity of matter to think and on common sense caught his eye, as did Priestley's analysis of human action, but he dismissed both volumes for their lack of distinct reasoning and philosophical precision.

Following up Priestley's glowing references to David Hartley, Reid read parts of Hartley's *Observations on man* the next day, and made an extensive summary of a number of propositions from the first and third chapters of Part One. Two topics interested him most: the relationship between vibrations in the brain and ideas in the mind, and Hartley's views on scientific method and logic. Regarding the former, it is not surprising that Reid was critical of the correlation between vibrations and ideas posited by Hartley, for he had already posed the rhetorical question in the *Inquiry*, 'how can the images of sound, taste, smell, colour, figure, and all sensible qualities be made out of the vibrations of musical chords, or the undulations of animal spirits, or of aether?'[122] Concerning the methodology espoused by Hartley, Reid made no explicit comment in these reading notes, but his later writings show that he objected to the Englishman's hypotheticalism, and it is likely that Reid was similarly critical of Hartley's application of mathematical modes of reasoning to the empirical sciences. Reid no doubt thought that Hartley was guilty of the same mistake as Francis Hutcheson in suggesting that mathematical methods should be used in moral philosophy, and whereas Hartley recommended the use of the calculus of chances to help correct common prejudices and fallacious reasoning, Reid did not think that the calculus was akin to inductive reasoning and he seems to have been sceptical about its applicability to everyday life.[123] Moreover, Reid sharply distinguished between the forms of reasoning appropriate for the abstract sciences of mathematics and those appropriate for the empirical sciences, and between those used in natural philosophy and those employed in such branches of moral philosophy as ethics, jurisprudence and natural theology.

Reid's 1774 notes from Hartley's *Observations* also shed significant light on the evolution of his methodological ideas. On the basis of Reid's correspondence with Lord Kames and the *Essays on the intellectual powers of man*, Larry Laudan has argued that Reid developed his methodology largely in response to Hartley's hypotheticalism and etherial physiology.[124] However, if one examines all of the available evidence, a different story emerges. As a Regent at King's College Aberdeen, Reid warned against the use of hypotheses in his lectures on pneumatology and natural philosophy, and Reid's criticisms of the theory of ideas in his orations delivered at graduating ceremonies held at King's during the 1750s and early 1760s foreshadowed his censure of hypotheses in the *Inquiry*.[125] I have already noted above that Reid's attack on the theory of ideas led him to study the physiology of perception, and that his antihypotheticalism evolved in the context of this review of the physiological mechanisms used to explain the physical basis of vision and the other human senses. We have seen too that during the 1750s Reid was apparently worried by the heterodox implications of Buffon's theory of generation and cosmogony, and it is likely that Reid's engagement with the *Histoire naturelle* provided a further stimulus for his critique of the hypothetical method. Following his move to the Glasgow Chair of Moral Philosophy in 1764, Reid refined his objections to the use of hypotheses, and in his lectures from this period his distinctive interpretation of Newton's First Rule of Philosophizing appears for the first time. Lecture notes dating from 1765 show that when he enumerated the competing physiological theories of perception, he rejected them on the ground that they failed to meet the two desiderata for acceptable causal explanations, namely that the cause 'ought to be true, & not a meer Fiction or Bare Conjecture without Proof [and] Secondly it ought to be sufficient to produce the Effect assigned to it'.[126] Later, in 1768–9, he revised his lectures, and in so doing he sharpened his formulation of the *vera causa* principle and presented it as an elaboration of Newton's First Rule. By 1769, then, Reid had evolved the interpretation of Newton's First Rule which he subsequently published in the *Essays on the intellectual powers of man*. Yet Reid's notes from Hartley's *Observations* indicate that he was unfamiliar with this work prior to 1774, and hence that he had formulated his methodological doctrines, and in particular his anti-hypotheticalism, *prior to* his reading of Hartley.[127] Thus it would seem that Hartley's advocacy of the hypothetical method and an etherial physiology were not the catalysts for Reid's inductivist methodology or his critique of hypotheses, as Laudan would have it. Rather, Reid's methodology apparently developed within the context of his attack on Hume and the theory of ideas, his rejection of the physiological theories which he believed underpinned that theory, and his opposition to heterodox thinkers such as Buffon. The writings of Hartley and Priestley may have elicited significant restatements of Reid's methodological views, but the available evidence suggests that they did not precipitate any important revision or reformulation.

Priestley's *Examination* of Reid, Beattie and Oswald duly appeared in the early autumn of 1774, and he presented each of them with a copy of the work.[128] The *Examination* was soon the cause of some mirth north of the Tweed, for Lord Kames reported to William Creech at the beginning of October that 'Dr Reid is here [at Blairdrummond] whom I employ'd to read passages out of Priestly for the amusement of us all'.[129] Reid also discussed the *Examination* with his Glasgow colleagues, and he followed the reviews and the correspondence about it in periodicals such as the *London chronicle*. But for the moment he remained silent publicly.[130] As he later disingenuously wrote to Richard Price regarding Priestley: 'I had resolved from the beginning . . . *to give him no Disturbance*'. In this letter, Reid also said of his opponent:

> I confess that in his late Examination &c he seems to me very lame in Abstract Reasoning as well as in some other qualities of more Estimation. I have got no Light from him to atone for his Abuse.[131]

As the documents reproduced in this volume illustrate, Reid did not subsequently alter this assessment of Priestley's philosophical talents, nor did he ever forgive him for the insults and obloquy contained in the *Examination*.

In 1775 Priestley published his *Hartley's theory of the human mind*, with the aim of making Hartley's system 'more intelligible, and the study of it more inviting'.[132] Priestley prefaced this edited version of Part One of the *Observations* with three introductory essays in which he attempted to clarify and defend Hartley's physiology and associationist psychology. For the first time in print, he here suggested that all of our mental powers arose from the 'organical structure' of the brain. Rejecting Hartley's vestigial dualism, Priestley argued that researches in chemistry had revealed that matter was subject to more than merely mechanical laws, and that the newly discovered subtlety and complexity of the laws governing the material creation lent credence to Locke's claim that 'a capacity of thinking might be given to matter'.[133]

Roused by this reinterpretation of Hartley, Reid published a lengthy two-part anonymous review of Priestley's *Hartley's theory of the human mind* in the *Monthly review* for July 1775 and January 1776. Here he angrily denounced Priestley's materialist hypothesis and attacked his associationist epistemology. Although it cannot now be determined whether this review was published at Reid's initiative or whether he was invited to comment on Priestley's edition by the co-founder of the *Monthly*, William Rose, a much more extensive autograph version of Reid's text, entitled 'Miscelaneous Reflections on Priestley's Account of Hartleys Theory of the Human Mind', has survived, and is reproduced below.[134] By comparing the published review with this manuscript we can see that Reid's text was heavily edited by Rose and his associates at the *Monthly*, and that in the substantial section of the manuscript devoted to Hartley's *Observations* that was omitted from the printed version, Reid was highly critical of various aspects of Hartley's work.

As his remarks make clear, Reid had little time for Hartley's account of the operations of the human mind or for his necessitarianism. But what he took greatest exception to was Hartley's mixture of fact and hypothesis in the *Observations*. According to Reid, Hartley had become too enamoured of his own necessitarian hypothesis, and had unwittingly furthered the cause of materialism by failing to distinguish between proven propositions and unsubstantiated conjectures.[135] Reiterating a point made in his 1774 reading notes, Reid claimed that whereas the first proposition in the *Observations* (which stated that the brain and the nervous system are the instruments of sensation and motion) was substantiated by the facts, the second (which drew a strict correlation between brain states and ideas) was not. This second proposition, which Reid thought was 'the main pillar of [Hartley's] system', had led the Englishman dangerously close to materialism, and Reid surmised that Priestley had followed the same path, only to go further and dispense with the concept of mind altogether. Reid therefore chided Hartley for not having imitated Newton's example of clearly demarcating between inductively proven conclusions and conjectures, as well as for having recommended the method of hypothesis contrary to Newton's prohibitions. Moreover, in Reid's view these methodological lapses were all the more regrettable for Priestley had failed to distinguish between the empirically well-founded and the purely hypothetical components of Hartley's work. Consequently Priestley had dogmatically presented Hartley's theory of the mind as a system based firmly on facts, and this misplaced belief in the truth of Hartley's ideas in turn had contributed to Priestley's conversion to materialism.

Reid also derived little comfort from the contents of Priestley's introductory essays on the association of ideas and on complex and abstract ideas, as Priestley's account of human action seemed to him to be the same as 'the selfish System', and his theory of the mind disturbingly similar to that of heretical thinkers like Epicurus, Hobbes and Hume. In particular, he thought that Priestley's reduction of our ideas of reflection to those of sensation could be used to bolster materialism, as it had been in the past by Epicurus and Hobbes (MS 3061/9, 11). Reid was disturbed as well by Priestley's explanation of instincts in terms of automatic responses and the association of ideas, as this seemed to him to rob them of their natural theological significance (MS 3061/9, 9–11). Quite apart from these matters of principle, Reid complained of Priestley's editing practice, his ignorance of the history of philosophy, and his lack of ability as a metaphysician (MS 3061/9, 4–5, 14, 17–20). The 'Miscelaneous Reflections' and the *Monthly* review based on them thus show that even before the appearance of the *Disquisitions relating to matter and spirit* in 1777, Reid considered Priestley's materialism to be the product of illegitimate modes of philosophical investigation, and that he was disturbed by the heterodox implications of the Dissenter's position. Furthermore, they manifest Reid's utter disdain for his new antagonist south of the Tweed. For whereas Reid had earlier treated other opponents like Hume with respect, he now responded to Priestley in a far more abrasive and personal manner than he had hitherto employed.[136]

Unfortunately, the chronology of subsequent events becomes increasingly difficult to establish, since few of the relevant manuscripts are dated, and for the crucial years 1779 to 1794 there are apparently no minutes extant for the proceedings of the Glasgow Literary Society, in whose meetings Reid attacked Priestley's materialism and necessitarianism. At some point in the period from April 1778 to the spring of 1784, Reid developed his criticisms of Hartley's methodology in the manuscript now catalogued as MS 2131/3/III/23 and transcribed below, which seems to be an early draft of passages which appear in Reid's *Essays on the intellectual powers of man*. One somewhat suprising feature of the 'Miscelaneous Reflections' was that Reid did not utilize his *vera causa* reading of Newton's First Rule of Philosophizing to undermine Hartley's theory of vibrations and vibratiuncles, but we can see from this manuscript that he soon began to do so.[137] Furthermore, it is significant that in this manuscript, as well as in the 'Miscelaneous Reflections' and the later *Essays*, Reid was at pains to stress Hartley's 'modesty' and 'candour', as if to underline the contrast he saw between Hartley's positive moral attributes and those of Hartley's self-appointed disciple Priestley.[138]

Reid evidently read Priestley's *Disquisitions relating to matter and spirit* and the exchanges between Priestley and Richard Price published as *A free discussion of the doctrines of materialism and philosophical necessity* (1778) shortly after they appeared. Although one of Reid's sets of notes from the *Disquisitions* is undated and the other comes from as late as May 1788, we know that he must have read it soon after its publication for there is a reference to it in his natural theology lectures for 1779–80.[139] The evidence is more sketchy regarding his study of the *Free discussion*, since the extant notes from it are undated, though Reid's colleague Robert Findlay drew his attention to a review of the work in the *Monthly review* for March 1779.[140]

Following his retirement from active teaching in 1780, Reid prepared his pneumatology lectures for the press, and in doing so formulated his reply to Priestley's criticisms of the *Inquiry* and his objections to Priestley's necessitarianism which appeared in his two volumes of *Essays*. From the fragments of them which survive, we know that Reid delivered three discourses on Priestley's materialism to the Glasgow Literary Society, but it is unclear exactly when these were read. Given the circumstantial evidence available, there seem to be three possibilities. Reid's undated notes from Priestley's *Disquisitions* were most likely written between 1779 and 1781, and they may be a sign that he was preparing his discourses attacking Priestley's materialism for the Literary Society. If this is the case, then Reid may have read these three discourses in the period 1779–83, since it would appear that he presented a series of discourses on free will and necessity to the Society between 1784 and 1786. A suggestive piece of internal evidence from Reid's manuscripts also points to this dating; it is perhaps significant that in his papers dealing with Priestley's materialism he does not mention the second edition of the *Disquisitions* which was published in London in 1782.[141] Assuming, therefore, that Reid read the three discourses on Priestley between 1779 and 1783, then MS 2131/2/III/13 (transcribed below) probably dates from 1782–3, for Reid's opening remarks suggest that this discourse formed a coda to the first three. Since MS 2131/2/I/15 (also transcribed below) begins with the sentence 'Some time ago I had the Honour to read to this Society Some Observations on the System of Materialism advanced and defended by the Revd Dr Priestley', it is likely that Reid read this discourse at some point in the period 1787 to 1793. On this reconstruction of the evidence, Reid's 'Some Observations On the Modern System of Materialism' would seem to have been written largely in the early 1780s, although Reid may have returned to the manuscript to make further revisions.

Alternatively, Reid may have first written the 'Observations' in the early 1780s, and then used his initial draft as the basis for a series of discourses delivered at a later date. MS 3061/23 (transcribed below) is the introduction to his third Society discourse, and the manner in which the manuscript ends indicates that Reid already had the text of the remainder of his discourse to hand. Comparison with MS 3061/1/4 suggests that Reid went on to present material from chapter five of the 'Observations', because the last line of MS 3061/23 serves as a linking passage to the beginning of the third paragraph of chapter 5 in MS 3061/1/4. Having read his discourses to the Literary Society, Reid may have subsequently revised the text of the 'Observations' in the light of his oral presentations.

Lastly, Reid may have written the 'Observations' and given his series of three discourses on Priestley during the years 1788 to 1790. The existence of the fairly detailed set of notes from the *Disquisitions* dating from May 1788 strongly suggests that he was working on his criticisms of Priestley in earnest at this juncture, and it is significant that the methodological issues raised in the notes were also discussed in the first of his discourses.[142] Otherwise, it is difficult to reconcile the detail of this set of notes with an earlier dating for the 'Observations' and the discourses, unless we assume that Reid's animus towards Priestley was such that he would repeat arguments in the notes which he had already stated elsewhere.

Moreover, since Reid was busy revising his lectures for publication following his retirement from teaching in 1780, it may well be that in the period 1780–3 he discoursed to the Society on topics dealt with in his *Essays on the intellectual powers of man*, and that he only turned his attention to Priestley's materialism once he had published both this volume and the *Essays on the active powers of man*, which he completed in 1786 or 1787. Circumstantial evidence gives some support to this particular hypothesis because Reid's *Inquiry* was based on discourses read before the Aberdeen Philosophical Society and the *Essays on the active powers of man* similarly incorporated materials discussed in both Aberdeen and Glasgow. Furthermore, during the years 1765 to 1778 Reid had discoursed exclusively on pneumatological subjects before his Glasgow colleagues, and papers on such topics as memory and the imagination later found their way into the *Essays on the intellectual powers of man*. From what we know of Reid's working habits and his contributions to the proceedings of the Literary Society, then, it appears quite likely that he would have tested the arguments he planned to incorporate in his *Essays on the intellectual powers of man* at the meetings of the Society held between 1780 and 1783.[143]

Consequently, if Reid did indeed read the three discourses on Priestley in the years 1788 to 1790, MS 2131/2/III/13 would then date from 1790–1, providing we accept that this discourse was a coda to the first three. In so far as the first line of MS 2131/2/I/15 indicates that some time had elapsed between the delivery of this discourse and the others, on this reading of the evidence it could only date from 1793, because Reid spoke to the Literary Society in 1794 and 1795 about political utopias and muscular motion respectively, and he could well have discussed Euclid's *Elements* in 1792.[144] However, it must be admitted that the opening of MS 2131/2/I/15 constitutes something of an anomaly for this alternative reconstruction, as Reid's choice of phrase here suggests a fairly lengthy intervening period. In any case, if this reconstruction does hold true, then 'Some Observations On the Modern System of Materialism' must date from the 1790s and, therefore, rank as Reid's last major philosophical work.[145] Moreover, the fact that he meticulously edited his text and had a fair copy prepared suggests that he intended to publish the 'Observations', and if he was revising his manuscript in the 1790s it may well be that its publication was prevented by Reid's ill-health and eventual death in 1796.[146]

As I have indicated, 'Some Observations On the Modern System of Materialism' can be read as an attempt to dissociate Newton's writings from Priestley's heterodox matter theory. According to Reid, what made Priestley's version of the materialist hypothesis distinctive was his claim that matter is an extended substance possessing inherent attractive and repulsive powers, rather than being inert, solid and impenetrable. For someone subscribing to a more conventional conception of matter as did Reid, Priestley's claim was controversial enough because it raised a host of conceptual and empirical issues. But what Reid found particularly perfidious was Priestley's self-proclaimed adherence to Newton's Rules of Philosophizing in order to justify his novel characterization of matter, since in Reid's view this stratagem gave the Dissenter's system a spurious air of legitimacy. Hence much of the 'Observations' is taken up with expounding the 'true' message of Newton's Rules, and exposing Priestley's misinterpretation of that message.[147]

Although Reid's lengthy methodological excursus in the 'Observations' is largely a summary of the essentials of the method he had long expounded in his lectures and writings, Priestley's presentation of Newton's Rules of Philosophizing did elicit what is perhaps the clearest and most systematic statement of Reid's own interpretation of the essence of Newton's methodology. Two features of Reid's discussion stand out as being of special importance. The first is Reid's careful definition of the scope of Newton's Rules. Like his Edinburgh counterpart Adam Ferguson, Reid here drew a sharp distinction between natural and 'civil or moral' phenomena, on the ground that the former were ultimately caused by God, whereas the latter resulted from the exercise of human will and power.[148] The upshot of this distinction for Reid was that whereas Newton's Rules were applicable within the domain of natural philosophy (which encompassed the study of natural phenomena and the laws governing the material system) and in pneumatology, they could not legitimately be employed in 'Morals, Jurisprudence, Natural Theology, . . . Mathematicks and Metaphysicks'.[149] Reid chastised Priestley for failing to recognize this distinction and for illicitly enlarging the scope of Newton's Rules, but in the 'Observations' Reid does not elaborate on why he thought this point was especially significant.

If we turn to his correspondence with Lord Kames, however, we see that Reid was hinting at an important question, namely whether all of the arguments surrounding Priestley's materialism could be resolved by appealing to Newton's methodological principles. In a well-known letter dated 16 December 1780, Reid outlined his views on hypotheses, Newton's ether and the respective limits of natural philosophy and natural theology or metaphysics. He also touched on the cognitive status of occasionalism, Leibniz's system of pre-ordained harmony and Priestley's materialist view of human nature, and he remarked that

> of all these systems about the efficient causes of the phenomena of nature, there is not one that, in my opinion, can be either proved or refuted from the principles of natural philosophy. They belong to metaphysics, and affect not natural philosophy, whether they be true or false.[150]

We see, then, that Reid believed that at least some of the issues raised by Priestley's materialism, such as whether matter is essentially active or passive, could not be adjudicated in terms of Newton's Rules of Philosophizing. In terms of Reid's map of human knowledge, therefore, his opponent's attempt to establish the truth of materialism using Newton's Rules alone was profoundly misguided, for Priestley had hopelessly confused questions of natural philosophy and metaphysics.

The second noteworthy feature of Reid's exposition of Newton's method is his blistering attack on Priestley's presentation of the *Regulae philosophandi* in the *Disquisitions*. As his reading notes from *circa* 1779–81 show, Reid was quick to spot apparent inaccuracies in Priestley's statement of the *Regulae*, and in the 'Observations' he adroitly turned what he thought were the inadequacies of the Dissenter's free translations to his own polemical ends.[151] While Reid translated Rule One as 'Of natural things no more causes ought to be admitted, than such as are both true and sufficient to explain their phenomena', Priestley's rendition read 'We are to *admit no more causes of things than are sufficient to explain appearances.*' Reid condemned this formulation of the First Rule on the ground that it licensed the use of hypotheses and conjectures. He contended that the historical record demonstrated that from the beginnings of philosophical speculation down to the time of Bacon, philosophers had been able to indulge their wit and formulate hypotheses with abandon, because theories were only required to be sufficient to explain the appearances. Bacon and Newton had thought otherwise, according to Reid, and the inductive method which they advocated (and which Newton had so triumphantly practised) was far more stringent than that which had hitherto prevailed in that it stipulated that any explanation of natural phenomena had to be *both* true and sufficient. In Reid's view, Newton had intended to combat the hypothetical method by demanding that causes be true, and by omitting that criterion Priestley had shown that he understood neither the spirit nor the letter of the First Rule.[152]

Reid likewise censured Priestley's wording of Rule Two. In Reid's translation it read 'Of natural effects of the same kind, the causes are the same', whereas Priestley rendered the Rule as '*to the same effects we must, as far as possible, assign the same causes*'. Reid complained that his opponent's addition of the phrase 'as far as possible' made the rule seem like an injunction to search for the simplest possible explanations. Reid had earlier warned against man's innate love of simplicity and its hazards in the *Inquiry*, and to him Priestley's interpolation mistakenly encouraged this propensity. Hence he cautioned 'that we be sure the effects be of the same kind before we assign them to the same cause'.[153]

Within the context of Priestley's attack on Reid's *Inquiry* in the *Examination*, their differences concerning the correct interpretation of Newton's Second Rule are highly significant. One of Priestley's main charges was that Reid had needlessly multiplied '*independent, arbitrary, instinctive principles*', and that as a result the Scottish philosopher's theory of the mind lacked 'the recommendation of that agreeable *simplicity*, which is so apparent in other parts of the constitution of nature'.[154] Presumably Reid regarded such an objection as being symptomatic of Priestley's misguided stress on '*simplifying* all appearances, and all causes' in philosophy, and Reid's remarks on Newton's Second Rule in the 'Observations' can be read as a methodological response to Priestley's criticisms of his epistemology.[155]

Among his opponent's sins of omission, none was more glaring for Reid than Priestley's failure to quote or comment upon Newton's Third Rule, which Reid translated as 'Qualities of Bodies which admit neither of increase nor diminution, and which are found to belong to all bodies on which we can make experiments, ought to be held as qualities of all bodies whatsoever'.[156] Here too, Reid was attempting to expose Priestley's avowed adherence to Newton's *Regulae* in the *Disquisitions* as disingenuous rhetoric, in so far as Priestley had attempted to ascertain the universal properties of matter without utilizing the one Rule of Philosophizing which governed such inductive generalizations. Reid insisted that such a Rule is necessary because our innate disposition to expect the future to be like the past leads us to make rash generalizations from the slightest evidence, and it is clear from subsequent sections of the 'Observations' that he thought that Priestley was guilty of drawing such 'lame and imperfect' inductions.[157] By revealing such inconsistencies between his opponent's procedures and Newton's principles and practice, Reid was again trying to show that Sir Isaac's methodology gave no succour to Priestley's materialism, and hence distance the Newtonian system (as he understood it) from the heterodoxies advanced by Priestley.[158]

In the sections of the 'Observations' dealing specifically with Priestley's conception of matter, Reid continued his attack on Priestley's method of philosophizing, and further underlined the disparity between Priestley's patterns of reasoning and Newton's *Regulae*. According to Reid, one notable instance of this disparity was Priestley's denial of the inertness of matter. Reid insisted that both experience and experiment had demonstrated that matter is inert, and hence that Newton's Third Rule sanctioned the generalization that inertia is a universal property of matter. Consequently Priestley's claim that matter lacks inertia was in Reid's view inconsistent with the message of this rule, as well as with all of the known empirical evidence.[159]

For Reid, another telling example of Priestley's methodological apostasy was his attribution of attractive and repulsive powers to matter itself, because Reid saw this as a glaring violation of Newton's injunction not to frame hypotheses. Reid drew a sharp distinction between those who followed Newton in eschewing causal explanations of gravity, and those philosophers like Priestley who indulged in conjectures in order to explain gravitational phenomena. Priestley's hypothesis that gravity is caused by a power of attraction essential to matter flew in the face of Newton's declaration 'hypotheses non fingo', and Reid contended that it contravened the First

Rule of Philosophizing since Priestley had not shown that an attractive power intrinsic to matter actually exists. Reid also deemed Priestley's hypothesis to be of no real explanatory value, because Newton had already sufficiently explained the phenomena in terms of his theory of gravitation without invoking a material cause for gravitational attraction.

Furthermore, Reid maintained that Priestley's remarks regarding Newton's ether speculations revealed that his opponent had no genuine understanding of Newton's basic attitude towards hypotheses. In the *Disquisitions* Priestley had indicated that Newton believed in the existence of an etherial medium which was the cause of a wide range of phenomena involving forces acting at a distance, but Reid protested that this was to misinterpret Newton's position on two counts. First, he thought that Priestley was wrong about the cognitive status of Newton's ether, because Newton had framed his remarks about an etherial medium in the form of Queries and had left the existence of such a medium a question open to resolution by future experimentation.[160] Secondly, Reid believed that Priestley had misconstrued the intent of Newton's ether hypothesis, for although Priestley seemed to think that Newton held that the ether itself was the material cause of the phenomena and hence regarded it as the *terminus ad quem* of all causal explanations, Reid claimed that Newton saw the ether as but another link in the causal chain leading to a wise and intelligent deity.[161] Reid stressed that Newton had no desire to explain the phenomena of nature in purely mechanical and material terms, and he countered Priestley by charging that it was the Dissenter's system rather than Newton's that led to a variant of Epicurean atheism. Newton's philosophy was free of any atheistic taint, Reid argued, because Newton had always insisted that matter was inactive and thus incapable of producing order and design, whereas Priestley had ascribed active powers to matter and hence allowed for the possibility that the beauty and symmetry of nature was the result of the action of matter alone. In his discussion of his opponent's matter theory, then, Reid effectively distanced Newton from Priestley by continuing to highlight the latter's methodological lapses, and by contrasting the orthodoxy of the Newtonian system with the heterodoxy of Priestley's materialism.

Reid had already noted in 1772 that Priestley lacked 'a philosophical Acumen in abstract Reasoning', and he evidently felt no inclination to revise this assessment when reading the *Disquisitions*. In the 'Observations', Reid catalogued a number of the Dissenter's failings in this regard, ranging from Priestley's lack of precision in his analysis of solidity and impenetrability, to his contradictory remarks about Newton's laws of motion and the ether.[162] More importantly, Reid maintained that Priestley was misguided in thinking that the question of the impenetrability of matter was an empirical one, because the issue was whether or not solidity was an *essential* property of matter, and hence not resolvable in terms of empirical data. Reid's argument here serves to illustrate the point made above that he did not think that materialism could be evaluated exclusively in empirical terms, and that he saw the need to invoke metaphysical and logical considerations in his deliberations.[163] Yet he was also more than willing to engage Priestley empirically, for Reid emphasized that the question of matter's inertia *was* a purely empirical one and, as we have seen, he insisted that experience and experiment demonstrate that inertia is a universal property of matter.[164]

Reid's personal dislike of his opponent also surfaces in an interesting way in the 'Observations'. Various passages in this work imply that Reid felt that Priestley's critical attitude towards Newton was highly presumptuous. For example, Reid rebuked Priestley for having implied that Newton subscribed to the idea that matter itself could act on other matter at a distance, and recalled that Newton had dismissed this idea as absurd in his letters to Richard Bentley.[165] Reid also made the ironical comment that Newton had lacked the sagacity to see that the laws of motion and the inertia of matter could be derived from an intrinsic power of repulsion in matter, and that it was left to Priestley to make this important discovery. We have seen too that Reid reprimanded his opponent for having suggested that Newton freely indulged in hypotheses, not least because this seemed to Reid to lessen Newton's stature as a natural philosopher. Finally, having detailed the numerous logical flaws in Priestley's arguments in the preceeding sections of the 'Observations', Reid concluded with an ironical twist, for he wrote that if Priestley's matter theory were true, then it would follow that Newton must have blundered in his reasoning, since the two of them appealed to the same facts and the same rules of philosophizing. Given the highly negative thrust of Reid's critique one could hardly assume the truth of Priestley's materialism, and the reader is thus left to infer that it was not Newton who had reasoned 'very ill'. Furthermore, Reid's invitation to Priestley to identify the fallacies in Newton's reasoning reads more like a challenge than a disinterested request for enlightenment. Hence Reid's reverential attitude towards Sir Isaac Newton prompted him to defend Newton's reputation in the 'Observations' against what he perceived as the slights and the misleading interpretations of Priestley, and there is little doubt that Reid judged Priestley guilty of hubris in setting himself up as Newton's philosophical rival. Consequently, in the 'Observations', Reid assumed the mantle of Samuel Clarke and acted as the defender of the true Newtonian faith, while once again casting Priestley in the role of the unfaithful disciple.[166]

One curious feature of the 'Observations' is Reid's virtual silence about Priestley's heretical religious views, for we have seen that in the 'Miscelaneous Reflections' Reid commented on the Dissenter's doctrinal principles and he probably also did so before the Glasgow Literary Society. In the fragment of the third of his discourses on Priestley published below, Reid referred explicitly to the Dissenter's Socinianism, and he echoed his earlier remarks in the 'Miscelaneous Reflections' concerning the incompatibility between materialism and Christianity.[167] Yet when he came to revise his discourses and draft the 'Observations', Reid toned down his references and contented himself with brief allusions to Priestley's anti-trinitarianism, although it must be said that Reid repaid the Dissenter an old debt when he suggested that Catholics could use the notion of the penetrability of matter to underwrite the doctrine of transubstantiation and that Trinitarians could similarly appeal to this notion in order to explicate the concept of consubstantiation.[168] More serious was Reid's accusation that Priestley's materialism led to atheism, but even here Reid simply stated this objection without labouring the point for polemical ends.[169] The lack of explicit, sustained discussion of theological and religious questions is thus striking, but if Reid intended to see the 'Observations' through the press, then the relatively restrained style of the final version probably reflects Reid's decision to limit his criticisms to philosophical matters and to avoid engaging Priestley in a broader polemic about religious issues. Having once been pilloried by Priestley's pugnacious pen, it may well be that Reid wanted to restrict potential debate as much as possible to metaphysics, and to leave more contentious topics aside.

Following up his review of Priestley's matter theory, Reid devoted two further Glasgow Literary Society discourses to refuting the fundamental tenets of materialism. In the first of these discourses (MS 2131/2/III/13), he was primarily concerned to rebut materialism by illustrating the role played by active, immaterial powers in the natural order. Here we see why Reid was so preoccupied with Priestley's denial of the inertness of matter, for it is clear that Reid's arguments for the existence of immaterial powers and of an intelligent first cause hinged on the premise that matter is inert. When speaking of the laws governing gravitation, magnetism and the other active powers first catalogued by Newton, Reid put the argument most succinctly: 'In all these Laws of Nature Active power is constantly exerted; There is no such Power in Matter; therefore, it must be in some Being that is immaterial'.[170] Hence Priestley's theory of matter struck at the foundation of the standard arguments deployed in Newtonian natural theology to demonstrate the existence of God and of immaterial powers subordinate to Him. It is scarcely surprising, then, that Reid insisted on the passivity of inanimate and 'animated' matter, and that he rejected the notion of an active ether as the cause of gravitation and the other attractive and repulsive powers observable in nature.[171]

Whereas materialists had tended to blur the distinction between the inanimate and animate realms of nature, in his penultimate discourse on materialism Reid countered this metaphysical move by maintaining a strict division between the two spheres. He argued that in the inanimate realm a given configuration of matter was no more than the sum of its parts, but that in the animate realm this was not the case, since an 'animated' body possessed a property not to be found in its parts, namely a unity or wholeness which we recognize as 'life'. While he admitted that the notion of life is obscure and ill-defined because we merely perceive the effects of life rather than the cause, he was adamant that this unity was the result of inanimate matter being united to an immaterial principle. According to Reid, the principle of life has three crucial characteristics. First, this principle must be indivisible, for Reid thought that it was inconceivable that it could have parts or be divisible. To him, this was an evident truth of reason which could not be overturned by such anomalous facts as the manner in which Trembley's polyp reproduced itself. The second attribute of the principle of life was its power to preserve the organization of living bodies. He defined the organization of the body in terms of the combination of its constituent material elements, and he believed that while the body is alive the principle of life protects it from the destructive effects of air, heat and moisture, as well as from the action of the gastric juices in the case of animals. Significantly, Reid stated that it is impossible for us to know how vegetables and animals first come to be organized. That he deemed the origins of organization to be unknowable may be related to the strategy sketched above in Section 1, of precluding materialist accounts of generation by drawing epistemological boundaries around key phenomena such as the physico-chemical processes involved in the formation of vegetable and animal bodies. However, it may be that he declared the origins of organization to be outwith the bounds of human knowledge because he found it difficult to reconcile the evidence about the chemical construction of vegetables and animals with his theory of organized atoms.[172]

The third feature of the principle of life Reid identified was its power to carry out the vital functions of living beings. He claimed that in so far as vegetables and animals were able to nourish themselves, reproduce their kind, expel what is noxious to the body, overcome disease, heal wounds and regenerate lost parts, there must be an immaterial power active in the animal oeconomy, and this he equated with the principle of life. By ascribing all of the vital functions to such an immaterial agent, Reid could therefore deny (as he does in this discourse) that the terms 'stimuli' and 'irritability' designate efficient causes of physiological phenomena, just as he denied that the term 'gravity' refers to the cause of gravitational effects, and it may be that he was anxious to do so because the concept of irritability had been appropriated by materialists like La Mettrie.

Although Reid nowhere indicated the sources for his theory of the principle of life, it would seem that he had drawn on the works of Stahl, Whytt and John Hunter, and it is likely that he had also been influenced by the ideas of his cousin and ex-colleague, John Gregory, who had begun speculating on the physiological functions of a 'vital principle' during his student days in Edinburgh in the

1740s.[173] While Reid was critical of aspects of Stahl's physiology, his attribution of a healing power and a power of maintaining the organization of the body is reminiscent of Stahl's definition of the functions of the *anima*.[174] Whytt's view that the body is a system 'whose functions are all owing to the power and agency of an immaterial sentient principle to which it is united, and by which every part of it is animated and put in motion' no doubt influenced Reid as well.[175] Finally, Reid's remarks on the power of digestion indicate that he was familiar with at least some of the papers in which John Hunter outlined his conception of a 'living principle'.[176]

Beyond defending the existence of immaterial beings, Reid was extremely reluctant in this discourse to elaborate on their specific nature in any detail. He was quite content to declare his ignorance about the differences between the immaterial principles operative in the inanimate and animate realms, and he refused to speculate on the fate of life-giving immaterial principles after the death of the vegetables and animals to which they had been united. However, Reid did insist that not all immaterial beings are endowed with thought, and he rejected the Cartesian view that thinking was an essential property of immaterial beings. He stressed that the immaterial principles active in the inanimate and vegetable kingdoms were devoid of thought, and that the principle of life in animals was unique in having that power.[177]

Such questions aside, Reid believed that the existence of principles of life in the vegetable and animal kingdoms strengthened the rational proof drawn from natural theology that there exists an immaterial soul in man. For if the role of vital principles throughout the whole animate creation was acknowledged, then this proof could be seen to rest on 'the whole Analogy of Nature'. He disagreed with those who were unwilling to admit for theological reasons that any animate being other than man was united to an immaterial agent, since he felt that the arguments for the existence of the human soul could be equally well applied to vegetables and animals if we allowed them life and activity. Presumably, he held that by denying that other living things have a kind of immaterial soul, Cartesians as well as some Christian apologists had unwittingly promoted the cause of heterodoxy by supplying materialists with arguments which could be used to bring into doubt the existence of an immaterial soul in man. Hence for Reid the defence of orthodoxy was better served by accepting that vegetables and animals were composed of both matter and an immaterial vital principle. He was careful to state, however, that the Butlerian analogy he employed in no way lessened the distinction between man and the inferior animals, as man's distinguishing feature was his rational and moral nature, which was 'the Image of his Maker'.[178]

The conclusion of this discourse is of particular interest, because Reid here briefly discusses Locke's suggestion that thought could be superadded to matter.[179] He may have addressed himself to Locke's thesis simply because it had inspired a number of materialists in the eighteenth century, but given the fact that Priestley had referred to Locke's conjecture in the prefatory essays to his edition of Hartley's *Observations*, it is more likely that Priestley's reference prompted Reid to consider Locke's text. Certain that Locke was not himself a materialist (and perhaps anxious to clear him of the charge), Reid explained that the great philosopher entertained his hypothesis because no one had convincingly demonstrated that the essential qualities of matter and thought are incompatible. Without such a demonstration the superaddition hypothesis remained a logical possibility, and Reid implied that Locke was enough of a metaphysician to have understood this. But Reid maintained that the advantage enjoyed by materialists was short-lived, because the requisite demonstration of the incompatibility of matter and thought had been essayed first by Samuel Clarke and then by the noted German mathematician and natural philosopher Leonhard Euler.[180] In Reid's view, therefore, both metaphysical argument and the empirical study of the Book of Nature disproved the materialist system and vindicated the existence of active immaterial agents.

The surviving fragment of Reid's second discourse, which deals with the question of the immateriality of the soul, covers much the same ground as his previous discourse and tells us little about what additional considerations (if any) he brought to bear on the problem which he addressed. Perhaps the most noteworthy features of this fragment are Reid's discussion of the relations between soul and body and his apologetic use of contemporary chemical theory. For Reid, the investigation of chemical phenomena revealed the operation of immaterial principles in the inanimate kingdom of matter, and highlighted our ignorance of both the number of composite bodies in nature and the manner in which they are compounded.

Thus he noted that although water had been assumed to be a simple elemental substance since antiquity, recent discoveries had shown that it was in fact composed of two kinds of heterogeneous elastic airs.[181] More generally, Reid affirmed that the science of chemistry had disclosed that the majority of observable bodies were composites, and he contended that this made the claim that man is similarly compounded of body and mind all the more plausible. Furthermore, he argued that since chemistry showed that the manner in which different substances were united together was mysterious, it was not surprising that the union between body and mind appeared to be inexplicable. Hence he used the findings of chemistry to defuse one of the major objections to dualism which Priestley himself had restated in the *Disquisitions*.[182] There is thus little doubt that Reid's remarks on the union of soul and body were prompted by Priestley's claim that the union of two such categorically distinct substances is unintelligible and, while it is true that his appeal to chemical evidence reflects to some extent his interest in the development of that science in the 1780s and 1790s, it would also seem that his arguments were directed specifically at Priestley, who was, after all, one of Britain's leading chemists as well as being an outspoken materialist.[183] Consequently, from what we know of Reid's last discourse on materialism it can justly be considered as the final chapter in his battle with Joseph Priestley, for his defence of the immateriality of the soul in this discourse is in effect a direct response to the Dissenter's metaphysical (as opposed to his historical) criticisms of dualism.[184] Clearly, Reid's obsession with his opponent was as strong as ever at the end of his life, and one wonders if Reid ever succeeded in exorcizing the spectre of Priestley's materialism to his own satisfaction.

By way of conclusion, one final question should be addressed, namely, were Reid's discourses on Priestley and materialism, and his 'Observations' in any way conditioned by the emergence of radical materialism and religious heterodoxy in France after the Revolution of 1789? It is tempting to see such a connection. Reid was caught up in the drama unfolding across the Channel and, like many in Britain, having initially been a supporter, he turned against the Revolution with the onset of the Terror in 1793.[185] Moreover, it is highly unlikely that he would have ignored Priestley's vigorous advocacy of the French cause. Yet we are here confronted with serious problems of chronology and a chronic lack of evidence. For we have seen that Reid may have penned much of his polemic against Priestley *prior to* 1789, which would mean that the broader concerns informing his attack had nothing to do with the events in France. Even if we accept that his discourses date mainly from the late 1780s and early 1790s, however, we must bear in mind that Reid probably began his examination of Priestley's *Disquisitions* in 1788, so that his critique of Priestley cannot have been linked originally to fears of rampant disbelief in French society. It may well be that Reid was prompted to rewrite the 'Observations' and compose his final discourses against materialism under the influence of growing fears about the course of the Revolution, but evidence to suggest such a connection is conspicuously lacking. It is difficult to believe that Reid would not have made some allusion to the situation in France, if this was indeed the source of his anxieties about Priestley's system and about materialism more generally in the 1790s. Since his papers are silent on this issue, a degree of scepticism concerning the possible connections between Reid's critique of Priestley and the rise of materialism in revolutionary France is in order.

We should also beware of any facile interpretation of this episode which would seek straightforwardly to relate the respective metaphysical positions of Reid and Priestley to differing socio-political orientations.[186] It is a striking fact that Reid and Priestley shared common social and political ground, even though they differed so dramatically over materialism (and necessitarianism). Both men were advocates of religious toleration. Both were opponents of slavery. Both were proponents of enlightened philanthropy, and both were Whiggish in their politics.[187] Yet their agreement in this sphere did not translate into a consensus over basic epistemological or metaphysical principles.[188] If we are to understand the dynamics of the Reid–Priestley affair, we need to look closely at the reasons which first prompted Priestley to launch his attack on the writings of Reid, James Beattie and James Oswald, for after Priestley fired his first salvo, Reid was arguably driven simply by a desire to defend both himself and the philosophical principles he held dear.

Whereas Joseph Priestley is sometimes seen as one of the leading spokesmen for the English Dissenting community as a whole, it must be said that not all Dissenters approved of the polemics he wrote on their behalf. In 1770, for example, Priestley was criticized by William Enfield, who in that year became the Tutor in Belles Lettres and Rector at the Warrington Academy.[189] Enfield argued in his *Remarks on several late publications relative to the dissenters; in a letter to Dr. Priestley. By a dissenter*, that the acerbic tone of Priestley's pamphlets on the legal disabilities suffered by the Dissenters and other related issues ill served their cause, and in the course of defending his *Remarks* against Priestley, he cited with approval the first volume of James Oswald's *An appeal to common sense in behalf of religion*, which had appeared in 1766. At this juncture Priestley had not read Oswald's work, for he wrote in response to Enfield that:

> When you talked of *self-evident and primary truths*, you should have explained your terms; for it was certainly very hazardous, to trust my understanding of you to my having read any particular book. The treatise you mention I have not yet seen, but I shall immediately procure it upon your recommendation.

Thus it seems that Priestley was prompted to read Oswald by his exchanges with Enfield, and it is significant that the merits of common-sense epistemology should have become an issue within the context of this debate, since Priestley's subsequent reading of Oswald was probably coloured to some extent by the fact that Enfield had deployed the *Appeal* against him.[190]

Another likely reason why Priestley was disposed to attack Reid and his Scottish compatriots has to do with the rise and rapid fall of Lord Bute. Bute's brief term as George III's first minister unleashed a wave of hostility towards the Scots among all ranks of society throughout England. Bute was especially disliked by Priestley and his political allies because they saw him as a corrupt and Machiavellian politician bent on illegitimately increasing the powers of the Crown, and as the figure responsible for what they believed were Britain's misguided policies regarding the American colonists. Many Englishmen harboured similar sentiments about the Scots more generally. Men such as Lord Shelburne (who later became Priestley's patron) held that 'the generality of Scotch . . . had no regard to truth whatsoever', and it was commonly believed that the Scots were all closet Catholics and that they remained sympathetic to the Stuarts and hence to absolutism.[191] The continuities between this xenophobic image and Priestley's later attack on

Reid, Beattie and Oswald is marked. In the *Examination*, Priestley charged that their epistemological principles gave succour to Catholicism, and that the doctrines of Common Sense would allow 'politicians . . . [to] venture once more to thunder out upon us their exploded doctrines of passive obedience and non-resistance'.[192] Such passages indicate that Priestley's perception of the Scottish philosophical triumvirate was structured by the events of the 1760s, and that he saw their epistemology as potentially sanctioning the priestcraft of both the papists and the ecclesiastical hierarchy of the Anglican Church, as well as the despotic political designs of George III and Lord Bute. Priestley's attack, in turn, seems to have inspired the scurrilous preface to the English translation of Claude Buffier's *Traité des premières veritez et de la source de nos jugements* issued by Priestley's publisher Joseph Johnson in 1780. Here too Reid, Beattie and Oswald were the targets of unrestrained xenophobia, and the anonymous translator dismissed them collectively as a group of 'Northern Book-makers' who had 'purloined' their ideas 'unacknowledged' from Buffier.[193] In the circles in which Priestley moved there was thus a strong current of anti-Scottish prejudice, which arguably found expression in the *Examination*.

Priestley's anxieties about the implications of Common Sense philosophy were made all the more intense by the circumstances surrounding the attempt begun in 1772 by London Dissenting ministers to obtain relief from subscribing to the Thirty-nine Articles. In the period in which Priestley published his *Institutes* and prepared the *Examination*, the London ministers seeking such relief were petitioning Parliament to alter the Toleration Act of 1689. Although Priestley was not directly involved in the ministers' deliberations until 1775, his support for the cause was well known.[194] The years 1770 to 1773 also saw James Beattie's rise to fashion among the London literati, with his *Essay on truth* gaining him a state pension from George III and the applause of the Archbishop of York along with other leading Anglican churchmen.[195] Moreover, circumstantial evidence suggests that Reid's *Inquiry* was enjoying wide currency in the metropolis and elsewhere, and we have seen that the writings of James Oswald were gaining an appreciative audience within the Dissenting community itself.[196] As the *Examination* makes abundantly clear, Priestley was deeply disturbed by the popularity which the works of his Scottish opponents were then enjoying,[197] and he firmly believed that the appeal to common sense was antithetical to rational religion. Taken in conjunction, these points suggest that Priestley feared that if Common Sense philosophy were to gain widespread acceptance at Court and within the upper echelons of the Anglican hierarchy, as well as with the public at large, then the Dissenters' campaign for relief from subscription would be jeopardized. An epistemology which (to him) celebrated subjective sentiments and feelings as the ultimate standards of truth, could, he seems to have thought, be used to justify the most arbitrary principles and policies, and perhaps be mobilized against his Dissenting brethren who were campaigning for relief from subscribing to the Thirty-nine Articles.[198] Furthermore, his fears about 'scotch principles and politics' must have been reinforced by the fact that the campaign for relief was opposed not only by the Anglican hierarchy but also by a number of prominent Scottish Presbyterian ministers in London, who voiced their disagreement in a petition presented to Parliament.[199] Once we look at the particular historical conjuncture in which Priestley wrote the *Examination*, we can see that his motivation was largely to further the political cause of Dissent by refuting a potentially dangerous philosophical resource for its opponents. It was for this complex of reasons, then, that in the years 1773 and 1774 Rational Dissent came into conflict with Scottish Common Sense.[200]

NOTES

1. Stewart, p. 18; compare Fraser, p. 35.
2. Stewart, pp. 14–23, esp. p. 19; compare Fraser, pp. 36–41, and McCosh, p. 201.
3. On Reid's friendship with Grant see Fraser, p. 35. Reid was a member of the Gordon's Mill Farming Club, which was founded in 1758 by a coterie of improvers drawn from King's College and the Aberdeen area; for further details see Wood, pp. 116–19, and J. H. Smith, *The Gordon's Mill Farming Club 1758–1764* (Edinburgh and London, 1962).
4. Reid's note on the discussion in the Philosophical Club is dealt with below. Given what Reid had to say about the religious utility of natural history in his course at King's College and his use of natural historical evidence in his natural theology lectures at Glasgow, it is extremely unlikely that his botanizing at New Machar was simply recreational.
5. With the curriculum reforms of 1753, regents at King's were required to lecture on natural and civil history; see P. B. Wood, *The Aberdeen*

Enlightenment: the arts curriculum in the eighteenth century (Aberdeen, 1993), ch. 3. Reid probably included some natural historical materials in his lectures prior to the new regulations.

6. AUL MSS 2131/3/II/12 (Réaumur – dated 1753); 3/II/15 (Tournefort); and 3/II/17 (Du Hamel de Monceau – dated 1761). For another set of notes related to Tournefort see MS 2131/6/v/14a.
7. MSS 2131/8/v/1, 1r–v; 6/v/13; 6/v/21, 1r. Unfortunately Reid left no indication of what texts he used for his lectures on the anatomy of animals or those on comparative anatomy.
8. MSS 2131/8/v/1, 1r–v; 6/v/1, 1v; 7/II/17, 1r. Reid's interest in plant physiology carried over into the Aberdeen Philosophical Society, where he proposed the following question for discussion in 1758: 'Whether some Part of that Food of Plants which is contained in the Air, is not absorbed by the Earth, and in the Form of a Watery Fluid conveyed into the Vessels of Plants? And whether anything can enter into the Vessels of Plants that is not perfectly soluble in Water?'; see *The minutes of the Aberdeen Philosophical Society 1758–1773*, ed. H. L. Ulman (Aberdeen, 1990), p. 189, and MS 37, 226v, for notes related to this question by David Skene. Reid's question (along with one proposed by his kinsman John Gregory in the same year) was prompted largely by utilitarian considerations, and well illustrates the impetus given to the pursuit of natural history by the improving ethos. Reid discussed a number of topics related to agricultural improvement in his lectures on the natural history of vegetables; see especially MSS 2131/8/v/1, 1r; and 6/v/21, 1r–v.
9. MSS 2131/6/v/4, 2r, 3r; 6/v/5, 1r, 4v, 5v; 6/v/13; 6/v/21; 6/v/22–22a; 7/II/17. A substantial section of Reid's lectures devoted to the 'histories' of various chemical substances is to be found in 'The Chemical History of Salts', MS 2343.
10. MS 2131/6/IV/1, headed 'Elements of Natural History', and MS 2131/6/v/10a, headed 'Natural History 1753'.
11. MSS 482, 44 (an entry dated 13 August 1764) and 56 (dated 20 August [?]), and MS 2131/6/v/1.
12. Some of Reid's letters to David Skene are published in *Works*, pp. 39–50. The originals of these letters have recently been rediscovered; see K. G. Kitagawa, '"Cadgers are ay speaking of Crooksadles": the rediscovered letters of Thomas Reid to Drs Andrew and David Skene', *Studies on Voltaire and the eighteenth century*, 314 (1993), 207–29.
13. For evidence of their collaboration see Reid to Joseph Black, 17 January 1773, in the Joseph Black Correspondence, EUL MS Gen. 873/1/59–60, and AUL MS 2343, which incorporates information supplied by George Skene. For his part, George Skene did not rate his talents as an analyst highly; see Skene to Joseph Black, 25 March 1773, in EUL MS Gen. 873/1/55–6, 1v–2r. In 1775, George Skene took over his father's chair in

civil and natural history at Marischal. For evidence of Reid's continued interest in the analytic art while at Glasgow see AUL MS 2131/3/I/14, 2r, which contains reading notes on a portable laboratory taken from the English translation of Cronstedt's *Mineralogy* published in 1770.

14. For Buffon's attack on taxonomy see especially George Louis Le Clerc, Comte de Buffon, *Histoire naturelle, générale et particulière, avec la description du cabinet du roi*, 44 vols (Paris, 1749–1804), I, 8–24. Reid's most detailed discussion of the epistemological and ontological foundations of the taxonomic enterprise occurs in the *Intellectual powers*, pp. 431–71. I have dealt with the points raised in this paragraph at greater length in 'Buffon's reception in Scotland: the Aberdeen connection', *Annals of science*, 44 (1987), 169–90, esp. pp. 171–3, 188–90.

15. Buffon, *Histoire naturelle*, I, 53–60; below, MS 2131/8/II/14, 1r–v, and Reid to James Gregory, undated, in *Works*, pp. 76–9, quote at p. 77.

16. See below, MS 2131/3/II/14, 2r. For Reid's censure of the world-makers in his natural theology lectures see 'Notes', v, 105, 172–4, and for similar remarks see *Inquiry*, p. 4. Buffon was attacked for his view of the age of the earth in a review of the *Histoire naturelle* published in the *Nouvelles ecclesiastiques* for 6 February 1750; see *From natural history to the history of nature: readings from Buffon and his critics*, ed. and trans. J. Lyon and P. R. Sloan (Notre Dame and London, 1981), pp. 243–4. For Buffon's dismissal of the deluge see Buffon, *Histoire naturelle*, I, 78–9, 98–9, 131–2; the last passage cited here would have been especially contentious. On the materialistic and atheistic implications of Buffon's cosmogony (and of the *Histoire naturelle* more generally) see G. Wattles, 'Buffon, d'Alembert and materialist atheism', *Studies on Voltaire and the eighteenth century*, 266 (1989), 285–341. The role of Buffon's theory of the earth in the rise of materialism and scientific naturalism in the eighteenth century is discussed in A. Vartanian, *Diderot and Descartes: a study of scientific naturalism in the Enlightenment* (Princeton, 1953), pp. 110–16.

17. See below, MS 2131/3/II/14, 3v (emphasis added). For Buffon's discussion of organic matter see especially *Histoire naturelle*, II, 33, 37–41, 43–53, 420–6. On Buffon's theory of generation see *inter alia* S. A. Roe, *Matter, life and generation: eighteenth-century embryology and the Haller–Wolff debate* (Cambridge, 1981), pp. 15–18; J. Roger, *Les Sciences de la vie dans la pensée française du XVIIIe siècle: la génération des animaux de Descartes à l'Encyclopédie*, 2nd edn (Paris, 1971), pp. 542–58; id., *Buffon: un philosophe au jardin du roi* (Paris, 1989), chs 9–10; C. Castellani, 'The problem of generation in Bonnet and in Buffon: a critical comparison', in *Science, medicine, and society in the Renaissance: essays to honor Walter Pagel*, ed. A. G. Debus, 2 vols (New York, 1972), I, 265–88; P. J. Bowler, 'Bonnet and Buffon: theories of generation and the problem of species', *Journal of the history of biology*, 6 (1973), 259–81. Buffon's

observations on spermatozoa are discussed in C. Castellani, 'Spermatozoan biology from Leeuwenhoeck to Spallanzani', *Journal of the history of biology*, 6 (1973), 37–68, esp. pp. 45–54.

18. John Stewart, 'Some remarks on the laws of motion, and the *inertia* of matter', *Essays and observations, physical and literary, read before a society in Edinburgh, and published by them* (Edinburgh, 1754), 70–140, quote at p. 72. For Reid's view of matter and impressed forces see below, MS 2131/2/III/13, A recto–B1v, and MS 2131/2/I/15, 2v.
19. [Anthony Collins], *Reflections on Mr. Clark's second defence of his letter to Mr. Dodwell*, 2nd edn, corrected (London, 1711), p. 25.
20. *Intellectual powers*, p. 621.
21. See especially Buffon, *Histoire naturelle*, II, 33.
22. 'Notes', v, 98; compare *Intellectual powers*, p. 631. Buffon's Scottish translator William Smellie perceived Buffon in similar terms; see Count de Buffon, *Natural history, general and particular*, [trans. W. Smellie], 8 vols (Edinburgh, 1780), I, xii–xiii n.
23. Buffon, *Histoire naturelle*, II, 32–3.
24. Significantly, Reid seems to have thought Buffon's theory of generation was essentially hypothetical. It is difficult to believe that he did not have Buffon in mind when attacking unidentified theories of the generation of animals in the introduction to the *Inquiry*, p. 4.
25. See especially Buffon's 'Discours sur la nature des animaux', in Buffon, *Histoire naturelle*, IV, 3–110, esp. pp. 22–4, 29–30, 36–41, 55–61, 86–90, 103–4.
26. See MS 2131/6/v/12, 2v, and MS 2131/6/v/35, 1r–v. That Reid was uneasy with Buffon's position is suggested by the fact that the notes he made from volume III of the *Histoire* are taken up exclusively with Buffon's description of the differences between the faculties of man and animals.
27. Below, MS 2131/6/v/35, 1r.
28. MS 2131/4/I/29, 16 (original pagination).
29. MS 2131/4/I/29, 17; compare below, MS 3061/9, 1, where Reid cites Buffon's dissection when censuring Priestley's materialism. Reid's remarks indicate that he had read Buffon's article on the orang-utan included in volume XIV of the *Histoire naturelle*, which first appeared in 1766. On Buffon's account of the orang-utan see R. Wokler, 'Tyson and Buffon on the orang-utan', *Studies on Voltaire and the eighteenth-century*, 155 (1976), 2301–19.
30. *Intellectual powers*, p. 437; see also p. 468 where Reid says that species are fixed because vegetables and animals have the power 'of producing their like'.
31. *Works*, p. 53n.
32. MS 2131/6/I/17, 1r; Reid and his associates here echo Samuel Clarke; see *The Leibniz–Clarke correspondence*, ed. H. G. Alexander (Manchester, 1956), p. 117.

33. MS 2131/6/I/17, 1r. On the question of wastage see F. J. Cole, *Early theories of sexual generation* (Oxford, 1930), p. 63, and Roe, *Matter, life and generation*, p. 9.
34. MSS 2131/8/v/1, 1v; 7/II/17, 1r; and 6/v/10a, 1v.
35. See Ulman, *Minutes*, pp. 85, 190. Two versions of the abstract from the discussion of this question survive in MS 37, 188r, and MS 475, 284–9 (original pagination).
36. J. Farley, *The spontaneous generation controversy from Descartes to Oparin* (Baltimore and London, 1977), pp. 18–29; Vartanian, *Diderot and Descartes*, pp. 258–62. On Needham see S. A. Roe, 'John Turberville Needham and the generation of living organisms', *Isis*, 74 (1983), 159–84; id., 'Metaphysics and materialism: Needham's response to d'Holbach', *Studies on Voltaire and the eighteenth century*, 284 (1991), 309–42.

    Although Peter Bowler and Jacques Roger have argued that the concepts of preformation and pre-existence should be distinguished clearly, this distinction is not crucial to the argument I am advancing here. I have therefore used the term 'pre-existence' simply to denote the belief in the existence of the embryo prior to conception. See P. J. Bowler, 'Preformation and pre-existence in the seventeenth century: a brief analysis', *Journal of the history of biology*, 4 (1971), 221–44, esp. p. 222.
37. MS 37, 188r; MS 475, 285.
38. MS 2131/8/VII, 65 (original pagination).
39. MS 2131/4/I/29, 7–8. Reid here refers to the researches of Harvey, Hartsoeker, Leeuwenhoek and Fallopius.
40. This is certainly the message of the passage in the introduction to the *Inquiry*, p. 4, where Reid writes that 'All our curious theories of the formation of the earth, [and] of the generation of animals . . . so far as they go beyond a just induction from facts, are vanity and folly, no less than the vortices of Des Cartes, or the Archaeus of Paracelsus.'
41. Reid to David Skene, 14 September 1767 and 31 October 1767, in *Works*, pp. 48–9. While in Glasgow, Reid probably shared his interests in natural history with William Irvine and John Anderson. Irvine lectured on *materia medica* from 1766 to 1787 and, although Anderson was the Professor of Natural Philosophy, his surviving manuscripts demonstrate that he was a keen natural historian as well; see P. Wood, ' "Jolly Jack Phosphorous" in the Venice of the North; or, Who was John Anderson?', in *The Glasgow Enlightenment*, ed. R. B. Sher and A. Hook (Edinburgh, 1995), pp. 111–32.
42. On this debate see S. A. Roe, 'Needham's controversy with Spallanzani: can animals be produced from plants?', in *Lazzaro Spallanzani e la biologia del settecento: teorie, esperimenti, istituzioni scientifiche*, ed. W. Bernardi and A. la Vergata (Florence, 1982), 295–303.
43. Reid to Lord Kames, [1775], in *Works*, pp. 53–4, and MS 2131/3/III/3.

The second paragraph of Reid's letter indicates that he had earlier written to Kames about the propagation of plants, which explains why he had little to say about this in their subsequent correspondence. The published version of Reid's letter, along with Kames's *Gentleman farmer*, displays the conjunction of theoretical speculation with practical application characteristic of the style of natural history which they shared with the majority of their contemporaries. In Scotland, the links between natural history and economic improvement were forged at the turn of the eighteenth century by Sir Robert Sibbald and his contemporaries; see R. L. Emerson, 'Sir Robert Sibbald, Kt, the Royal Society of Scotland and the origins of the Scottish Enlightenment', *Annals of science*, 45 (1988), 41–72.

44. Reid to Kames, [1775], *Works*, pp. 53–4, quote at p. 53. In the letter sent to Kames, Reid did not speculate on when the organized atoms were formed, but in the draft version he wrote that we simply do not have any evidence which would determine whether organized atoms existed 'from the beginning of the world or from some later period' (MS 2131/3/III/3, 1r). Thus according to the distinction drawn by Bowler and Roger, Reid was neither a preformationist nor a believer in pre-existence at this stage in his career; see Bowler, 'Preformation and pre-existence', p. 222.

45. Reid to Kames, [1775], *Works*, pp. 53–4, quote at p. 53. Reid suggested that the organized atom required the 'proper degree of Moisture and heat' to unfold (MS 2131/3/III/3, 1v). See also MS 2131/2/III/13, B1v–2r, below.

46. MS 2131/3/III/3, 1r.

47. MS 2131/3/III/3, 1r; Reid to Kames, [1775], *Works*, pp. 53–4, esp. p. 54. In the letter and draft, Reid relied mainly on Harvey's *Exercitationes de generatione animalium* (1651) for his information on the development of the embryo. A note on the last page of the manuscript of the draft reads: 'All that I know of the Philosophy of Generation is grounded on the Facts observed By the immortal Harvey and Malpighi' (MS 2131/3/III/3, 2v). Roe points out that Malpighi's observations were commonly appropriated by proponents of pre-existence, even though he himself did not advocate this theory; see Roe, *Matter, life, and generation*, p. 6.

48. A. Vartanian, 'Trembley's polyp, La Mettrie, and eighteenth-century French materialism', *Journal of the history of ideas*, 11 (1950), 259–86; see also V. P. Dawson, 'Trembley, Bonnet, Réaumur and the issue of biological continuity', in *Studies in eighteenth-century culture*, ed. O. M. Brack, Jr., 13 (1984), 43–63; id., *Nature's enigma: the problem of the polyp in the letters of Bonnet, Trembley and Réaumur* (Philadelphia, 1987).

49. MS 2131/3/III/3, 1v.

50. MS 2131/3/III/3, 1v. The experiments are discussed in Roe, 'John Turberville Needham', pp. 161–3.

51. MS 2131/3/III/3, 1v; Reid to Kames, [1775], *Works*, pp. 53–4, esp. p. 54.
52. MS 2131/3/III/3, 1v. Reid had written to Richard Price in early 1775 on this point; see Reid to Price, 10 April 1775, in *The correspondence of Richard Price*, ed. D. O. Thomas and B. Peach, 3 vols (Durham, NC and Cardiff, 1983–94), I, 192–5.
53. MS 2131/3/III/3, 1v–2r; Reid to Kames, [1775], in *Works*, pp. 53–4, esp. p. 54.
54. MS 2131/3/III/3, 2r; *Leibniz–Clarke Correspondence*, pp. 23–4, 35, 53, 114–15.
55. John Walker to Lord Kames, 8 November 1775, SRO, MS GD24/1/571, 10r; compare Henry Home, Lord Kames, *The gentleman farmer; being an attempt to improve agriculture, by subjecting it to the test of rational principles*, 3rd edn (Edinburgh, 1788), pp. 437–8. The whole of the fifth appendix to this work is devoted to the 'Propagation of plants'. Kames's patronage helped Walker become the Regius Professor of Natural History and Keeper of the Museum in the University of Edinburgh in 1779; see S. A. Shapin, 'The Royal Society of Edinburgh: a study of the social context of Hanoverian science' (unpublished Ph.D. dissertation, University of Pennsylvania, 1971), pp. 152–4.
56. Kames, *Gentleman farmer*, p. 425 (italics added). In the mid-1750s Kames had been accused of materialism by John Stewart; see John Stewart, 'Some remarks on the laws of motion', *passim*. For an important discussion of the Kames–Stewart debate see M. Barfoot, 'James Gregory (1753–1821) and Scottish scientific metaphysics, 1750–1800' (unpublished Ph.D. dissertation, University of Edinburgh, 1983), ch. 2.
57. MS 2131/4/I/29, 8.
58. See below, MS 2131/3/I/25, 1v–3r and MS 2131/3/II/16, 1r.
59. GUL MS Murray 505, 54 (original pagination); see also Reid to Kames, 27 February 1778, in I. Ross (ed.), 'Unpublished letters of Thomas Reid to Lord Kames, 1762–1782', *Texas studies in literature and language*, 7 (1965), 17–65, esp. pp. 32–3.
60. This edition is listed in the biographical entry on Bonnet in the *Nouvelle biographie universelle*, 46 vols (Paris, 1852–66), V, 627–31, esp. p. 631. Bonnet's collected works were issued in various formats beginning in 1779.
61. The quotation on the title page of Priestley, *Disquisitions*, reads: 'Si quelqu'un demontreroit jamais, qu l'âme est materielle, loin de s'en alarmer, il faudroit admirer la puissance, qui auroit donné a la matiere la capacité de penser', which I translate as 'If anyone should ever demonstrate that the soul is material, far from being alarmed, one should admire the power that would have given matter the capacity to think.'
62. Kames, *Gentleman farmer*, pp. 421, 435; Kames referred to Bonnet's

Consideration sur les corps organisée (2 vols in 1 (Amsterdam, 1762). It is unclear whether Reid was familiar with this work; see his ambiguous comments below, MS 2131/3/II/16, 1r.

63. See below, MS 2131/3/II/16, 1r–2v. It should be noted that Bonnet paid tribute to, and deployed, Haller's theory of irritability in pt. x, ch. 7 of the *Palingénésie*. Reid's most extensive discussion of instinct occurs in his *Active powers*, pp. 103–17; see also p. 120 where he states that both instinct and habit are 'parts of our original constitution' for which 'we can assign no cause . . . but the will of him who made us'.
64. For similar claims of nescience see 'Notes', v, 54, 59.
65. See below MS 2131/2/III/13, B1v–2r; Reid to Kames, [1775], MS 2131/3/III/3, 1v. Bonnet had been deeply troubled by the metaphysical implications of Trembley's discoveries in the 1740s; see Dawson, 'Trembley, Bonnet, and Réaumur', pp. 55–60.
66. Reid to Kames, 16 December 1780, *Works*, pp. 56–60, esp. p. 56.
67. P. L. Farber, 'Research traditions in eighteenth-century natural history', in Bernardi and La Vergata, *Lazzaro Spallanzani*, 397–403, quote at pp. 398–9.
68. See Wood, 'Buffon's reception', pp. 188–9. I have investigated the use of natural historical methods in the study of humankind in eighteenth-century Scotland in 'The natural history of man in the Scottish Enlightenment', *History of science*, 28 (1990), 89–123. Reid intended to include what he called 'Select Historys of particular Species' in his natural history lectures, which suggests that he would have had some sympathy for Buffon's call for exact descriptions of each species of plant and animal in the *Histoire naturelle*; see MS 2131/8/v/1, 1v.
69. For a humorous treatment of the faddishness of theories of muscular motion see Tobias Smollett, *The adventures of Roderick Random*, ed. P.-G. Boucé (Oxford, 1981), p. 27.
70. MS 2131/6/I/17, 1r. On Baxter see R. K. French, 'Ether and physiology', in *Conceptions of ether: studies in the history of ether theories 1740–1900*, ed. G. N. Cantor and M. J. S. Hodge (Cambridge, 1981), 111–34, esp. p. 121.
71. French points out that Newtonian physiologists faced serious difficulties in solving the problem of the source of motion in the human body because of their emphasis on the inertness of matter; French, 'Ether and physiology', pp. 120–4.
72. MS 2131/7/II/17, 1r; see also his 'Scheme of a Course of Philosophy', MS 2131/8/v/1, 1v, where Reid lists 'The Muscles and Organs of Voluntary Motion' as a lecture subject.
73. Robert Whytt, *The works of Robert Whytt, M.D.* (Edinburgh, 1768), p. 171. On Whytt's physiology see R. K. French, *Robert Whytt, the soul, and medicine* (London, 1969); C. J. Lawrence, 'Medicine as culture:

Edinburgh and the Scottish Enlightenment' (unpublished Ph.D. dissertation, University of London, 1984), ch. 4; J. P. Wright, 'Metaphysics and physiology: mind, body and the animal economy in eighteenth century Scotland', in *Studies in the philosophy of the Scottish Enlightenment*, ed. M. A. Stewart (Oxford, 1990), 251–301, esp. pp. 276–92.
74. Whytt, *Works*, pp. 98, 110, 142–3, 145, 147–8, 153.
75. Whytt, *Works*, pp. v–vi, 2, 6, 31, 152, 172.
76. Whytt, *Works*, p. 208.
77. 'Of the Involuntary Motions of Animals', below, MS 2131/7/II/19, 1r. This manuscript almost certainly dates from the 1750s.
78. See below MS 2131/7/II/2, 1r–v. Reid's worries about the term 'stimulus' were perhaps prompted by the reservations expressed by Albrecht von Haller regarding Whytt's ideas; on the debate between Haller and Whytt see French, *Robert Whytt*, ch. 6, and S. A. Roe, '*Anatomia animata*: the Newtonian physiology of Albrecht von Haller', in *Transformation and tradition in the sciences: essays in honor of I. Bernard Cohen*, ed. E. Mendelsohn (Cambridge, 1984), 273–300. On Stahl's physiology see L. J. Rather, 'G. E. Stahl's psychological physiology', *Bulletin of the history of medicine*, 35 (1961), 37–49; L. S. King, 'Stahl and Hoffman: a study in eighteenth-century animism', *Journal of the history of medicine and allied sciences*, 19 (1964), 118–30; F. Duchesneau, *La physiologie des lumières: empirisme, modèles et théories* (The Hague, Boston and London, 1981), ch. 1; J. Geyer-Kordesch, 'Passions and the ghost in the machine: or What Not to Ask about science in seventeenth- and eighteenth-century Germany', in *The medical revolution of the seventeenth century*, ed. R. French and A. Wear (Cambridge, 1989), 145–63. Reid's third suggestion in this manuscript corresponds to Whytt's explanation of our lack of consciousness of the stimuli of vital motions.
79. *Intellectual powers*, pp. 96–104.
80. Reid's interpretation has been criticized in J. W. Yolton, *Perceptual acquaintance from Descartes to Reid* (Minneapolis and Oxford, 1984), passim.
81. Descartes' physiology is outlined in *Traité de l'homme*, *La Dioptrique*, *Le Discours sur la methode*, and *Passions de l'ame*.
82. René Descartes, *Treatise on man*, in Descartes, I, 106.
83. René Descartes, *The Passions of the soul*, in Descartes, I, 340; compare *Intellectual powers*, pp. 97–8.
84. On Descartes' concept of 'ideas' see N. Kemp Smith, *New studies in the philosophy of Descartes: Descartes as a pioneer* (London, 1952), pp. 146–60; E. Michael and F. S. Michael, 'Corporeal ideas in seventeenth-century psychology', *Journal of the history of ideas*, 50 (1989), 31–48, esp. pp. 31–6.
85. Nicolas Malebranche, *The search after truth*, trans. T. M. Lennon and

P. J. Olscamp (Columbus, Ohio, 1980), esp. pp. 17, 49–50, 87–136, 140–1, 152–3, 162–8, 337–9, 402–3, 449–50, 502–9.
86. Malebranche, *Search*, p. 102.
87. John Locke, *An essay concerning human understanding*, ed. P. H. Nidditch (Oxford, 1975), I. i. 2.
88. Locke, *Essay*, II. viii. 4, x. 5, xxxiii. 6.
89. *Intellectual powers*, pp. 98–104, 145.
90. It should be noted that Reid believed that physiologists like William Briggs and even Newton had been misled in their theorizing by assuming the existence of ideas; *Inquiry*, pp. 352–3, 357–63. The physiological basis of the theory of ideas in the eighteenth century merits further examination. On this question see K. M. Figlio, 'Theories of perception and the physiology of mind in the late eighteenth century', *History of science*, 12 (1975), 177–212; and J. P. Wright, *The sceptical realism of David Hume* (Manchester, 1983), pp. 15, 68–71, 204–21, where Wright explores the physiological presuppositions of Hume's version of the theory of ideas.
91. *Inquiry*, pp. 352–7; MS 2131/4/II/1, 20–2. In the *Inquiry*, Reid explicitly criticized William Briggs, and he probably also had in mind Descartes, Newton and lesser figures such as his contact in London, the physician Alexander Stuart. Stuart was an exponent of the theory of animal spirits, as can be seen in his *Three lectures on muscular motion*, which appeared as a supplement to the *Phil. Trans.*, 40 (1738), especially p. xliii. It is said that Reid's kinsman and colleague John Gregory was similarly critical of William Cullen's etherial physiology; see R. E. Schofield, *Mechanism and materialism: British natural philosophy in an age of reason* (Princeton, 1970), p. 207. I have as yet been unable to find evidence to confirm this, but Gregory did warn against the abuse of hypotheses in his *Lectures on the duties and qualifications of a physician*, new enlarged edn (London, 1772), pp. 149–50; on Gregory's physiology and philosophy see Lawrence, 'Medicine as culture', ch. 7.
92. Schofield, *Mechanism and materialism*, ch. 9; T. M. Brown, 'From mechanism to vitalism in eighteenth-century English physiology', *Journal of the history of biology*, 7 (1974), 179–216.
93. *Inquiry*, p. 472. Compare Gregory, *Lectures*, p. 154, where he paraphrases Reid. Gregory also spoke here of 'an internal principle, which feels, which thinks, and which seems to be the origin of animal motions', whose operations were not reducible to 'the laws of matter'.
94. William Porterfield, 'An Essay concerning the motions of our eyes', *Medical essays and observations*, 3 (1735), 160–261, 4 (1737), 124–294; id., *A treatise on the eye, the manner and phaenomena of vision*, 2 vols (Edinburgh, 1759); Whytt, *Works*, pp. 161–3, quote at p. 163; for an account of the differences between Whytt and Porterfield see Wright, 'Metaphysics and physiology', pp. 286–9.

95. For Reid on Stahlian physiology see *Thomas Reid's lectures on the fine arts*, ed. P. Kivy (The Hague, 1973), pp. 25–6, and for Reid on Porterfield see *Inquiry*, especially pp. 261–2, 341–7.
96. 'Notes', v, 69–72. Despite Reid's assertion to the contrary, the claim that the nerves were akin to hollow tubes did have some evidence in its favour; see Edwin Clarke, 'The doctrine of the hollow nerve in the seventeenth and eighteenth centuries', in *Medicine, science, and culture: historical essays in honor of Owsei Temkin*, ed. L. G. Stevenson and R. P. Multhauf (Baltimore, 1968), 123–41.
97. As his 'Scheme of a Course of Philosophy' from 1752 indicates, Reid thought that poetry and painting illustrated 'the Effect of Body & Mind upon each other' (MS 2131/8/v/1, 2r). Reid was influenced here by John Gregory; see Gregory's 'A Proposall for a Medicall Society Written anno 1743', AUL MS 2206–45, 4 (original pagination). Reid seems to have derived his fourfold classification of theories of the mind–body relation in his lectures from Herman Boerhaave; for Boerhaave's analysis see *Dr. Boerhaave's academical lectures on the theory of physic. Being a genuine translation of his institutes and explanatory comment, collated and adjusted to each other, as they were dictated to his students at the University of Leyden*, 2 vols (London, 1742–3), I, 69. I owe this point to John Wright.
98. Kivy, *Lectures*, p. 25.
99. Kivy, *Lectures*, p. 25. The text is here somewhat garbled, as is often the case with student notes.
100. See Reid's discussion of habit in the *Active powers*, pp. 117–21.
101. Kivy, *Lectures*, p. 25; see also p. 28. It should be noted that the question of how the mind can effect the body was linked to the much broader metaphysical issue of how immaterial principles can act on matter. It must be said that Reid did not resolve this problem satisfactorily, and that his later writings on materialism (which are discussed below) beg this issue. In a similar vein John Gregory argued that it was pointless to enquire into the nature of the union between the soul and the body, since this was probably beyond the limits of our understanding and was of no conceivable practical consequence. Instead, he recommended that physicians should study the laws governing the union of body and soul; Gregory, *Lectures*, pp. 109–11, 165–6. Like Reid, Whytt believed that although we do not know how the will effects voluntary motions, experience none the less convinces us that this is the case; Whytt, *Works*, pp. 2, 122.
102. Kivy, *Lectures*, p. 26.
103. See below MS 2131/2/III/13, B2v–3r.
104. Reid alludes to Haller's theory in 'Some Observations On the Modern System of Materialism', below, MS 3061/1/4, 33–4. On Haller see French 'Ether and physiology', p. 131. For La Mettrie see Julien Offray de La

Mettrie, *Man a machine*, trans. G. C. Bussey and M. W. Calkins (LaSalle, 1912), pp. 129–31.
105. See below MS 3061/2, 25.
106. MS 3061/2, 17; compare 'Notes', v, 71–2.
107. MS 3061/2, 4.
108. Reid's terminology here is strikingly reminiscent of Whytt. In the *Essay* Whytt wrote: 'The immediate cause of muscular contraction, which ... appears to be lodged in the brain and nerves, I chuse to distinguish by the terms of the *power* or *influence of the nerves*' (Whytt, *Works*, p. 6).
109. MS 3061/2, 18–24. On our ignorance of the nature of the nervous power compare Whytt, *Works*, p. 172. Similarly sceptical remarks are to be found in *Inquiry*, p. 238; 'Notes', v, 72; 'Some Observations On the Modern System of Materialism', below, MS 3061/1/4, 34; and the undated fragment transcribed below MS 2131/7/II/19 in which Reid states that physiologists had been totally unsuccessful in accounting for the vital motions of the body. Nevertheless, he says that there is an 'Active principle' in animals which is the origin of motion.
110. MS 3061/2, 19.
111. MS 3061/2, 20; W. C. Walker, 'Animal electricity before Galvani', *Annals of science*, 2 (1937), 84–113, esp. pp. 101–3.
112. MS 3061/2, 20–1.
113. Reid's view of occasionalism is discussed in C. J. McCracken, *Malebranche and British philosophy* (Oxford, 1983), pp. 308–11; see also p. 147 for Locke's claim that Malebranche's occasionalism implied that man was not morally responsible for his actions.
114. That is, Priestley's Newton is more the speculative natural philosopher, whereas Reid's Newton is the austere inductivist and critic of the hypothetical method. On the question of differing images of Newton in the eighteenth century see the seminal paper by H. Guerlac, 'Where the statue stood: divergent loyalties to Newton in the eighteenth century', in *Essays and papers in the history of modern science* (Baltimore and London, 1977), 131–45; see also G. Buchdahl, *The image of Newton and Locke in the Age of Reason* (London and New York, 1961); and R. E. Schofield, 'An evolutionary taxonomy of eighteenth-century Newtonianisms', in *Studies in eighteenth-century culture*, ed. R. Runte, 7 (1978), 175–90. For pertinent reflections on this issue with reference to Priestley see S. Schaffer, 'Priestley's questions: an historiographic survey', *History of science*, 22 (1984), 151–83, esp. p. 172.
115. Stewart, esp. pp. 96–115, 169; Fraser, pp. 94–101. It should be noted that Stewart outlines the incompatibility between Reid's philosophy and materialism, without going into specific details about Reid's writings on Priestley's system. Fraser at least tries to say something concrete about Reid's reply to Priestley.

116. McCosh, pp. 206, 473.
117. See for example J. H. Faurot, 'Reid's answer to Joseph Priestley', *Journal of the history of ideas*, 39 (1978), 285–92.
118. J. W. Yolton, *Thinking matter: materialism in eighteenth-century Britain* (Minneapolis and Oxford, 1983), ch. 6. Reid's manuscripts transcribed below underline Yolton's historiographical point (pp. xi–xii) that eighteenth-century British materialism cannot be properly understood without reference to the natural philosophical and physiological ideas of the period.
119. See MSS 2131/4/II/4, 1r, 2131/4/I/29, 16 and 'Notes', IV, lecture 71 (this volume is unpaginated). As Baird's notes make clear, at some point in the 1770s Reid seems also to have read d'Holbach's *Système de la nature* (1770). Reid's logic lectures show him to have been familiar with Humphrey Ditton's *A discourse concerning the resurrection of Jesus Christ . . . together with an appendix concerning the impossible production of thought from matter and motion . . .* (London, 1712); see 'A System of Logic, Taught at Aberdeen 1763, by Dr Thomas Reid', EUL MS Dk.3.2, 84. For further documentation of Reid's knowledge of the materialist canon, see my 'Thomas Reid's critique of Joseph Priestley: context and chronology', *Man and nature: proceedings of the Canadian Society for Eighteenth-Century Studies*, ed. D. H. Jory and C. Stewart-Robertson, 4 (1985), 29–45. On materialism in eighteenth-century Britain more generally see Yolton, *Thinking matter, passim*.
120. Priestley, *Institutes*, III, v. Priestley read Reid's *Inquiry* soon after it was published and later referred to it in *The history and present state of discoveries relating to vision, light, and colours* (London, 1772; Millwood, NY, 1978), pp. 658, 665; compare Priestley, *Examination*, pp. vii–viii. Reid's reading notes from Priestley's *History and present state of electricity* (1767), taken in August 1768, survive; see MS 2131/3/I/9.
121. Priestley, *Examination*, pp. 346–71; see also Priestley's letter to Caleb Rotherham, 31 May 1774, in *A scientific autobiography of Joseph Priestley, 1733–1804: selected scientific correspondence, with commentary*, ed. R. E. Schofield (Cambridge, Mass. and London, 1966), 145–6, esp. p. 146.
122. *Inquiry*, p. 355.
123. See below MS 3061/1/4, 15–16. Compare David Hartley, *Observations on man, his frame, his duty, and his expectations*, 2 vols (London, 1749; New York, 1971), I, 335–9, with MS 2131/5/II/49, 1r–v.
124. L. Laudan, 'The medium and its message: a study of some philosophical controversies about ether', in Cantor and Hodge, *Conceptions of ether*, pp. 157–85, esp. p. 171.
125. This evidence is surveyed in my 'Reid on hypotheses and the ether: a

reassessment', in *The philosophy of Thomas Reid*, ed. M. Dalgarno and E. Matthews (Dordrecht, 1989), 433–46.
126. MS 2131/4/II/1, 20 (original pagination).
127. Although it could be argued that it is still possible that Reid read Hartley prior to 1774, none of the extant evidence suggests this. I have not found a single reference to Hartley in any of Reid's papers or in the student lecture notes which date from before 1774. Significantly, only George Baird's notes from 1779–80 contain an explicit reference to Hartley's physiological ether; see 'Notes', I, 39–42. Even if we admit that he may have read Hartley in the 1750s or 1760s, the evidence still indicates that Reid only engaged with Hartley's ideas seriously after reading Priestley's *Institutes* and his edition of the *Observations*. It should also be remembered that Priestley was largely instrumental in popularizing Hartley's work, so that there is no reason to believe that Reid would necessarily have come across it prior to 1774. The *Observations* did not reach a second edition until 1791, while Priestley's version appeared in a second edition in 1790. Furthermore, the extent to which Hartley was a figure known to the Scottish literati can be gauged by the fact that Lord Monboddo had not heard of Hartley until the late 1770s. Writing to Richard Price, Monboddo said of Hartley that he had 'never heard so much as the name till I was last in London [in 1779]' (Lord Monboddo to Richard Price, 15 September 1780, in W. Knight, *Lord Monboddo and some of his contemporaries* (London, 1900), 124–37 quote at p. 126). Monboddo apparently first learned of Hartley through the controversy provoked by Priestley's works.
128. Priestley, *Examination*, pp. iii–v
129. Lord Kames to William Creech, 4 October 1774, SRO, Dalguise Muniments, GD38/2/19.
130. For an extensive undated abstract of the *Examination* see MS 3061/15. In a letter to Richard Price, Reid mentioned discussions with his Glasgow associates concerning Priestley's work, and the correspondence about it in the *London chronicle* (Reid to Price, [10 April 1775], in *Price correspondence*, I, 195–8, esp. p. 198). For the exchanges concerning the *Examination* see *London chronicle*, 36 (1774), pp. 425, 532. Reid later told Lord Kames that 'Beattie is to answer Priestley with profound Silence' (Reid to Lord Kames, 7 September 1775, SRO, The Abercairney Papers, GD/24/1/569, 85r). Reid also comments on aspects of Priestley's materialism in this letter, and his remarks (conflated with part of another letter) are printed in *Works*, p. 52.
131. Reid to Richard Price, [10 April 1775], in *Price correspondence*, I, 195–8, esp. p. 198. To Priestley's credit, in *A free discussion* he noted that Reid was an able opponent, and he later regretted the tone of the *Examination*; see Priestley, *A free discussion*, p. 189, and the *Autobiography of Joseph Priestley*, ed. J. Lindsay (Bath, 1970), p. 113.

132. Priestley, *Hartley*, p. 369.
133. Priestley, *Hartley*, pp. xix–xxi.
134. See MS 3061/9. In the manuscript list which seems to have been compiled in the nineteenth century (now catalogued as MS 3061/26), it is claimed that the 'Miscelaneous Reflections' were read before the Glasgow Literary Society, and this claim was repeated by Reid's biographer A. Campbell Fraser in Fraser, p. 115. However, there is no record of this in the extant minutes of the Society for this period. What little evidence survives suggests that Reid was in fairly close contact with William Rose, who reviewed Reid's *Inquiry* in the *Monthly* in 1764 and later found a publisher for Reid's two volumes of *Essays*; see the *Monthly review*, 30 (1764), 358–78, 31 (1764), 1–21; and Reid to James Gregory, 14 March 1784, in *Works*, pp. 63–4, esp. p. 64. Rose's son Samuel was a student at Glasgow from 1783 to 1787, and on at least one occasion he delivered papers from Reid to his father; see Reid to James Gregory, ibid., and W. Innes Addison, *The matriculation albums of the University of Glasgow: from 1728–1858* (Glasgow, 1913), no. 4268.
135. One of Reid's standard criticisms of the hypothetical method was that favourite hypotheses invariably led philosophers astray. In his lectures on optics at King's College Aberdeen, for example, Reid referred to Fermat's and Leibniz's derivation of the sine law of refraction in order 'to show that the most ingenious men when they trust to Hypothesis ... have only the chance of going wrong in a more ingenious way' (AUL MS K. 160, 263).
136. It would seem that Rose and/or his editors tempered the tone of the published review, as someone excised the *ad hominem* attacks on Priestley. It is significant that throughout the 'Miscelaneous Reflections', Reid indicated that, for all his faults, Hartley was a far better metaphysician than Priestley, and he also praised Hume's account of the association of ideas.
137. George Baird's notes from 1779–80 show that Reid was using the *vera causa* principle to attack both Hartley and Priestley in his lectures; see 'Notes', I, 39–42.
138. Reid singled out Priestley's 'manners' and 'candour' for comment when making notes on Priestley's *Examination*; see MS 3061/15, 4v. Reid also wrote to Richard Price that Priestley should have 'taken a lesson of Meekness good Manners and Candour' from 'Dr Hartly his Paragon' and Price (Reid to Richard Price, [10 April 1775], in *Price correspondence*, I, 195–8, esp. p. 198). On the candour of Unitarians like Priestley and Priestley's candour see the perceptive remarks by Donald Davie in *A gathered church: the literature of the English dissenting interest, 1700–1930* (New York, 1978), pp. 140–1, and M. Fitzpatrick, 'Toleration and Truth', *Enlightenment and dissent*, 1 (1982), 3–31, esp. pp. 25–7.

139. See 'Notes', IV, lecture 71. Reid's reading notes are transcribed below. The undated notes in MS 3061/12 are to be found with a discussion of Bishop Butler which is dated 'Aug 22 1781' on 2r.
140. See MS 2131/3/III/22, 1v.
141. For a brief summary of the contents of Reid's initial series of discourses on Priestley's materialism see below MS 3061/23, 2r–3v. For the dating of Reid's discourses on free will and necessity see MS 3061/22, 1r and 2v. One further piece of circumstantial evidence points to the period of 1779 to 1783 as being the date of Reid's discourses. Reid's colleague John Anderson noted that the two of them discussed Priestley during the summer of 1782 while they were staying at Dumbarton; John Anderson, 'Dumbarton Castle', AL MS 33 (unpaginated).
142. Compare below MS 3061/14 and MS 3061/23, 3r.
143. For further documentation of the genesis of the two *Essays* and Reid's contributions to the Glasgow Literary Society see Wood, pp. 170–3, 176–87, and Appendix III.
144. Reid originally read his discourse on Euclid before the Aberdeen Philosophical Society in January 1762, and he later delivered a revised version in the Glasgow Literary Society; see Wood, pp. 215, 219 and the references cited there. Reid was formally excused from having to deliver discourses in 1794, presumably because of his failing health; GUL MS Gen. 4, 2. However, he did manage to give his discourses for 1794 and 1795, but his death in October 1796 prevented him from speaking before the Society in that year.
145. My discussion here of the dating of Reid's discourses on Priestley and the 'Observations', differs substantially from that offered in my 'Thomas Reid's critique of Joseph Priestley', pp. 38–9, and Wood, p. 324. I no longer consider my earlier reconstruction to be accurate.
146. It should be noted, however, that James Gregory reported that when asked by William Creech to publish on Priestley, Reid remarked 'What, Mr. Creech, would you have me wrestle with a chimney sweeper!'; quoted in Michael Barfoot, 'Priestley, Reid's Circle and the Third Organon of Human Reasoning', in *Science, medicine and dissent: Joseph Priestley (1733–1804)*, ed. R. G. W. Anderson and C. Lawrence (London, 1987), 81–9, esp. p. 84.
147. Reid's exposition of 'Newton's' method in his lectures, his public orations at King's, and in the *Inquiry* performed a similar function with respect to David Hume. Hume too appealed to Newton's *Regulae*, and the main thrust of Reid's attack on Hume's system was that it violated the rules of philosophizing set down by Newton; for Hume's reference to Newton's Rules see David Hume, *Enquiries concerning human understanding and concerning the principles of morals*, ed. L. A. Selby-Bigge and P. H. Nidditch, 3rd edn (Oxford, 1975), p. 204.

148. See below MS 3061/1/4, 13, 1–16; compare Adam Ferguson's distinction between physical and moral laws in his *Institutes of moral philosophy*, 3rd edn, enlarged (Edinburgh, 1785), pp. 4–7, 79–83, 135–8.
149. Reid's insistence that Newton's Rules can be employed in pneumatology is problematic because he elsewhere states that 'our contemplative powers are under the guidance and direction of the active'; see *Intellectual powers*, p. 67. Given this 'fact', it would seem that the phenomena of the human mind are largely moral phenomena, and hence outwith the scope of Newton's *Regulae*.
150. Reid to Lord Kames, 16 December 1780, in *Works*, pp. 56–60, quote at p. 59. Compare here Reid's remarks in the introductory section of his King's College natural philosophy lectures, where he argues that the question of the passivity or activity of matter lies within the province of pneumatology, not natural philosophy; 'Natural Philosophy 1758', AUL MS K.160, 10.
151. The gist of what Reid has to say about Priestley's treatment of the first two Rules of Philosophizing in the 'Observations' appears in the notes on the *Disquisitions* dating from 1788; see below MS 3061/14, 1r, 2v.
152. MS 3061/1/4, 17–19; Priestley, *Disquisitions*, p. 2. In the Motte–Cajori translation the First Rule reads: 'We are to admit no more causes of natural things than such as are both true and sufficient to explain their appearances' (Newton, *Principia*, II, 398). Significantly, in 1788 Reid similarly noted that Boscovich and Benedetto Stay misstated Newton's First Rule, and omitted the truth (i.e. existence) requirement; see below, MS 2131/3/I/17, 5. On the context of Reid's 1788 reading notes from Stay and Boscovich see R. Olson, 'The reception of Boscovich's ideas in Scotland', *Isis*, 60 (1969), 91–103; and the suggestive remarks in M. Goldie, 'The Scottish Catholic Enlightenment', *Journal of British studies*, 30 (1991), 20–62, esp. pp. 55–7. Reid first learned of Boscovich's theory of matter in 1757 through reading a paper on optics by Thomas Melvill; see Wood, pp. 244–6 and MS 2131/3/I/10.
153. MS 3061/1/4, 19–20; *Inquiry*, pp. 470–3; Priestley, *Disquisitions*, p. 2. The Motte–Cajori translation reads: 'Therefore to the same natural effects we must, as far as possible, assign the same causes' (Newton, *Principia*, II, 398). Reid was evidently not aware of the fact that Newton had altered the wording of the Second Rule in the third edition of the *Principia* (1726), where Newton added the phrase 'as far as possible' (*quatenus fieri potest*) to the original formulation contained in the first two editions; on this change see A. Koyré, 'Newton's "regulae philosophandi"', in his *Newtonian studies* (Chicago, 1968), pp. 261–72, esp. p. 266. It would seem, therefore, that the methodological differences between Reid and Priestley were partly the result of their using different editions of the *Principia*.

154. Priestley, *Examination*, p. 6.
155. Priestley, *Examination*, p. 1.
156. MS 3061/1/4, 20–3. The Motte–Cajori translation reads: 'The qualities of bodies, which admit neither intensification nor remission of degrees, and which are found to belong to all bodies within the reach of our experiments, are to be esteemed the universal qualities of all bodies whatsoever' (Newton, *Principia*, II, 398).
157. On the issue of premature inductions see also *Inquiry*, pp. 70–1.
158. It is interesting that neither Reid nor Priestley make any reference to Newton's Fourth Rule, added to the third edition of the *Principia*. One of the few instances where Reid mentions Rule Four is in the introduction to his natural philosophy lectures; see MS K.160, 8.
159. MS 3061/1/4, 22. Reid presumably regarded Priestley's denial of the inertness of matter as an example of someone being misled by an hypothesis.
160. Reid here refers to Newton's *Opticks*, Queries 18–24 (these queries were added to the 1717–18 edition). On the vexed question of Newton's mature ether speculations see H. Guerlac, 'Newton's optical aether: his draft of a proposed addition to his *Opticks*', in his *Essays and papers*, pp. 120–30; J. E. McGuire, 'Force, active principles, and Newton's invisible realm', *Ambix*, 15 (1968), 154–208, esp. pp. 174–87; id., 'Neoplatonism and active principles: Newton and the *corpus hermeticum*', in R. Westman and J. E. McGuire, *Hermeticism and the scientific revolution* (Los Angeles, 1977), 95–142, esp. pp. 109–19.
161. Compare here Reid to Kames, 16 December 1780, *Works*, pp. 56–60, esp. pp. 57–8).
162. MS 3061/1/4, 26–32, 44–8.
163. In his King's College natural philosophy lectures, Reid affirmed that the solidity or impenetrability of matter was an 'axiom' of natural philosophy; see MS K.160, 9.
164. MS 3061/1/4, 24–5. Reid's comments on the status of the question of matter's solidity well illustrates the contrasting philosophical styles of the two men, in so far as Priestley's empirical arguments reflect his experimental expertise and orientation, whereas Reid's approach manifests his metaphysical preoccupations. Priestley's experimental practices are rightly emphasized in Schaffer, 'Priestley's questions', pp. 158–70.
165. See Newton's third letter to Bentley in *Isaac Newton's papers and letters on natural philosophy*, ed. I. B. Cohen (Cambridge, 1958), 302–3. In *A free discussion*, Richard Price similarly referred to Newton's letter when criticizing Priestley (Priestley, *A free discussion*, p. 29).
166. For Reid's various criticisms of Priestley see MS 3061/1/4, 48–50. As I have indicated, Reid employed much the same stratagem in the 'Miscelaneous Reflections' and the 'Observations', that is, accusing Priestley of

corrupting the doctrines of his masters, namely Hartley and Newton. John Stephens has argued that Richard Price likewise acted as a defender of the true Newtonian faith in his exchanges with Priestley (J. Stephens, 'Price, providence and the *Principia*', *Enlightenment and dissent*, 6 (1987), 77–93, esp. pp. 77–8).

167. MS 3061, 1r. Reid returned to such broader issues at the close of the first of his later discourses on materialism; see below MS 2131/2/III/13, B3r–4r.
168. MS 3061/1/4, 26. In the *Examination* Priestley had charged that the philosophy of common sense could be used by Catholic apologists; see Priestley, *Examination*, pp. 200–1.
169. MS 3061/1/4, 43, 47. For his part, Priestley had charged Reid, Beattie and Oswald with implicit atheism in *A free discussion* (Priestley, *A free discussion*, pp. xxvii–viii).
170. See below MS 2131/2/III/13, B1v.
171. MS 2131/2/III/13, Ar–v. Richard Price likewise stressed the central importance of inertia in *A free discussion*; see Priestley, *A free discussion*, pp. 3–4. For earlier formulations of the Newtonian position see Richard Bentley, *A confutation of atheism from the origin and frame of the world*, in *Newton's papers and letters*, 313–94, esp. pp. 340–4; Newton, *Opticks*, Query 31, pp. 397–402.
172. Reid had earlier told Lord Kames that it is impossible for us to know how the organization of organic beings comes about; see Reid to Kames, [1775], *Works*, pp. 53–4, esp. p. 54. Reid's interest in the chemical composition of vegetable and animal matter is evident in his undated manuscript on 'The Chemical History of Salts' (MS 2343) and in the lengthy reading notes he took in May 1789 from William Nicholson's translation of Fourcroy's *Elémens d'histoire naturelle et de chimie*, MS 2131/3/I/16, 26–30 (original pagination).
173. Gregory served as a Regent at King's College from 1746 to 1749, and returned as Professor of Medicine from 1755 to 1766. For his conception of the vital principle see 'A Proposall for a Medicall Society', 6. On Gregory's physiology more generally see Lawrence, 'Medicine as culture', ch. 7.
174. On Stahl's physiology, see the references cited above, n. 78.
175. Whytt, *Works*, p. 171. Reid's agnosticism about the precise nature of the immaterial agents active in the natural order would have accomodated Whytt's controversial notion of the sentient principle.
176. See especially John Hunter, 'On the digestion of the stomach after death', *Phil. trans.*, 62 (1773), 447–58.
177. See below MS 2131/2/III/13, A recto, B3r–v; compare C verso, where Reid argues that it is reasonable to infer that there are various intermediate orders of immaterial but unthinking beings on the ground that, from what we know of the hierarchy of nature, God would not have left the

enormous gap between brute matter and the lowest degree of intelligent being unfilled. It should be noted that in his last discourse on materialism, Reid pointed to one important difference between the immaterial principles operative in the inanimate and animate realms, namely that inanimate bodies are not united to the immaterial agents acting upon them, unlike vegetables and animals; see below MS 2131/2/I/15, 2v.

178. MS 2131/2/III/13, B3v. Robert Whytt had similarly criticized the Cartesians and 'some theological writers' (Whytt, *Works*, p. 153). Reid had probably first been alerted to the metaphysical dangers of the Cartesian beast machine in his philosophy classes with George Turnbull; see Turnbull's thesis *De scientiae naturalis cum philosphia morali conjunctione* (Aberdeen, 1723), p. 8, 'Annexa' 4.

179. Locke made this suggestion in the *Essay*, IV. iii. 6.

180. Reid refers here to Samuel Clarke's exchanges with Anthony Collins, and in particular Clarke's *A third defence of an argument made use of in a letter to Mr. Dodwell, to prove the immateriality and natural immortality of the soul. In a letter to the author of the reflections on Mr. Clarke's Second Defence, &c.*, in *The works of Samuel Clarke, D.D. Late Rector of St. James's Westminster*, 4 vols (London, 1738), III, 836–41. I have been unable to identify which work of Euler's Reid has in mind. Reid was familiar with at least some of Euler's writings on mechanics, though he does not mention a specific title; see MS 2131/3/I/8, 1v. Like Reid, Euler devoted much thought to the question of the solidity and impenetrability of matter, and it is unfortunate that Reid's references are so tantalizing yet so vague; for Euler's views on solidity and impenetrability see S. Gaukroger, 'The metaphysics of impenetrability: Euler's conception of force', *British journal for the history of science*, 15 (1982), 132–54.

181. See below MS 2131/2/I/15, 1v. On the discovery that water is a compound see J. R. Partington, *A history of chemistry*, 4 vols (London, 1961–70), III, 344–62. The discovery was announced by Lavoisier in 1783 and discussed in the Royal Society and the *Phil. trans.* in 1784.

182. Priestley, *Disquisitions*, pp. 60–74. That Priestley was led to adopt materialism because of the problem of mind–body interaction has been argued most recently in A. Tapper, 'The beginnings of Priestley's materialism', *Enlightenment and dissent*, 1 (1982), 73–81. However, Tapper does not explore how Priestley's electrical and chemical inquiries may have affected his matter theory. Priestley himself hinted that his work in chemistry had suggested that matter is active (Priestley, *Hartley*, p. xviii).

183. It should be noted that in a manuscript dating from 1766 Reid similarly appealed to the evidence of chemistry and natural philosophy to prove that the soul is immortal; see MS 2131/4/II/19, 2r. Although Reid made such an appeal in the 1760s, it still seems reasonable to assume that in

his later discourse he used evidence drawn from chemistry especially to combat Priestley. Reid's natural theological use of chemistry in this context provides a sharp contrast with the style of chemistry cultivated by William Cullen; see J. R. R. Christie, 'Ether and the science of chemistry: 1740–1790', in Cantor and Hodge, *Conceptions of ether*, 85–110, esp. pp. 93–4.

184. It must remain an open question whether Reid perceived the parallel between Priestley's historical account of the origins of the popular and philosophical concepts of the soul in the *Disquisitions* and the claim made by a number of Deists that the traditional conception of the soul was merely a heathen relic; see, for example, *The philosophical works of the late Right Honourable Henry St. John, Lord Viscount Bolingbroke*, 5 vols (London, 1754), I, 218 n.; and Priestley, *Disquisitions*, pp. 192–224, 241–347.

185. On Reid's reaction to the French Revolution see Wood, pp. 181–4.

186. For an example of the *kind* of approach I have in mind here see S. Shapin, 'Social uses of science', in *The ferment of knowledge: studies in the historiography of eighteenth-century science*, ed. G. S. Rousseau and R. Porter (Cambridge, 1980), 93–139, esp. pp. 111–24, 134–9. Shapin draws on the ideas of Mary Douglas in order to relate cosmologies to social structures; for an insightful and sensitive discussion of the problems inherent in such an enterprise see C. B. Wilde, 'Matter and spirit as natural symbols in eighteenth-century British natural philosophy', *British journal for the history of science*, 15 (1982), 99–131.

187. On Reid's social and political views see Wood, pp. 178–86, 188–92; Thomas Reid, *Practical ethics. Being lectures and papers on natural religion, self-government, natural jurisprudence, and the law of nations*, ed. K. Haakonssen (Princeton, 1990), pp. 70–85; K. G. Kitagawa, 'Not without the highest justice: the origins and development of Thomas Reid's political thought' (unpublished Ph.D. dissertation, University of Edinburgh, 1994).

188. The same could be said about Priestley and Richard Price, who represents an even clearer anomaly for the sort of approach recommended by Shapin.

189. Priestley had earlier served as Tutor in Languages and Belles Lettres at Warrington from 1761 to 1767. For details of Enfield's life and career see the biographical memoir by John Aikin included in Enfield's posthumously published *Sermons on practical subjects*, 3 vols (London, 1798), I, iii–xxvii. Enfield apparently had some connection with Glasgow, because he visited the Professor of Natural Philosophy, John Anderson, at Dumbarton Castle in the summer of 1782; see Anderson, 'Dumbarton Castle', AL MS 33.

190. [William Enfield], *A second letter to the Rev. Dr. Priestley* (n.p., [1770]), pp. 3–4; Joseph Priestley, *An answer to a second letter to Dr. Priestley*

([London, 1770]), p. 4; Priestley's *Answer* is dated 6 September 1770. In his initial reply to Enfield, Priestley accused him of laying 'the foundation of universal scepticism', and stated that '[o]f all the maxims of popery, the most dangerous is generally esteemed to be their keeping the common people in ignorance' (Joseph Priestley, *Letters to the author of remarks on several late publications relative to the dissenters, in a letter to Dr. Priestley* (London, 1770), pp. 59, 61). Priestley's rhetoric here is like that of the *Examination*.

191. For Shelburne's remark see J. Brewer, 'The misfortunes of Lord Bute: a case-study in eighteenth-century political argument and public opinion', *The historical journal*, 16 (1973), 3–43, esp. p. 20; Brewer discusses English perceptions of the Scots more generally on pp. 19–22. Linda Colley provides a useful perspective on the Scottophobia of the period in her *Britons: forging the nation 1707–1837* (New Haven, Conn. and London, 1992), ch. 3.

192. Priestley, *Examination*, p. 201. Priestley here develops a theme to be found in the writings of John Locke. In the *Essay* (I. iv. 24) Locke argued that the doctrine of innate ideas could be used as a means of social and political control.

193. *First truths, and the origin of our opinions, explained: with an enquiry into the sentiments of modern philosophers, relative to our primary ideas of things . . . to which is prefixed, a detection of the plagiarism, concealment, and ingratitude of the Doctors Reid, Beattie, and Oswald* (London, 1780), pp. vi–xii, xiv–xvii, xxiv, lvii–lix, quote at p. xv. Significantly, on pp. xvii and lviii the translator echoes Priestley's view that Reid was more of a philosopher than either Oswald or Beattie. It should be noted that Priestley himself drew attention to Buffier's writings in the first edition of his *Hartley's theory of the human mind*, although he did not accuse his Scottish opponents of plagiarism; Priestley, *Hartley*, p. 372.

194. On this episode see R. B. Barlow, *Citizenship and conscience: a study in the theory and practice of religious toleration in England during the eighteenth century* (Philadelphia, 1962), ch. 5; J. Stephens, 'The London ministers and subscription, 1772–1779', *Enlightenment and dissent*, 1 (1982), 43–71.

195. Beattie's success is chronicled in Sir William Forbes, *An account of the life and writings of James Beattie, LL.D.* (London, 1824), pp. 144–61. The Archbishop of York at this time was Robert Hay Drummond, who was the younger brother of the Earl of Kinnoul, whom Beattie knew personally; see Forbes, *Account*, p. 144. I thank Roger Emerson for pointing out this connection to me.

196. On the popularity of the *Inquiry* see Bennet Langton to James Boswell, 17 June 1773, in *The correspondence of James Boswell with certain members of the Club*, ed. C. N. Fifer (London, 1976), p. 30. It should be noted that the *Inquiry* had gone through three editions by 1770.

197. Priestley, *Examination*, pp. 200–2.
198. It is significant in this regard that Priestley judged that in the *Essay*, Beattie had 'approache[d] too near to the spirit of persecution', and remarked that 'I shall expect, after a summary process before the tribunal of his common sense, to be consigned to the disposal of his *friends of truth*, who may not be equally the friends and lovers of *mercy*' (Priestley, *Examination*, pp. 190, 192).
199. Barlow, *Citizenship and conscience*, pp. 183–8; Richard Price to the Earl of Shelburne, 23 February 1773, and Price to the Earl of Chatham, 11 March 1773, in *Price correspondence*, I, 155–7, 157–60, esp. pp. 157, 159. The phrase 'scotch principles and politics' occurs in a letter to Richard Price from the Virginian politician Arthur Lee, dated 20 April 1777, in *Price correspondence*, I, 253–5, on p. 253. The anti-Scottish prejudices of some of Priestley and Price's associates are graphically illustrated by Lee's remarks on p. 254 that 'in my opinion, it would require a People of more virtue than the world ever yet produced, or than human nature will admit of, to resist the contagion of scotch principles – to be united with Scotland, and not be undone. I mean as to its morals and public principles.'
200. For his part, Priestley believed that the Dissenting ministers failed in their efforts because of the despotic designs of the Court, the corrupt use of patronage, and the machinations of the Anglican bishops; see Joseph Priestley, *An address to Protestant Dissenters of all denominations, on the approaching election of members to parliament, with respect to the state of public liberty in general, and of American affairs in particular* (London, 1774), pp. 4–5.

The Manuscripts

PART ONE
Natural History

I

Histoire Naturelle Generale & Particuliere avec la Description Du Cabinet du Roi

Tom 1. 2d Edition a Paris 1750 par M. Buffon Intendant de Jardin Royale de plantes & M Daubenton. Garde & Demonstrateur de Cabinet d'Histoire Naturelle.

1 Discourse of the Manner of Studying Natural Historie pages 62. Buffon. In this Discourse he disparages our Modern Systems much as being almost wholly employed in Classing Dividing & Subdividing Natural Bodies; and neglecting the full & distinct Narration of Facts which he represents as the onely valuable part of Natural Historie in this he says the Ancients particularly Aristotle have vastly excelled the Moderns. Nature seems to have given no Countenance to any Nice division it is solely the work of the human Mind. Nature has made such a profusion of every kind as leaves no Gap betwixt one Genus and another nor even does she leave room for a line of division.

Linnaeus is severly censured for his System both of the Vegetable & animal Kingdoms.

We have here some Metaphysical Disquisitions about Truth which is divided into Mathematical & Physical The first he seems to disparage by representing it as purely a Creature of the Mind. The last Conclusion in a Mathematical Reasoning is identical with its andicedents & that with its Anticedent and so on, therefore Mathematical conclusions are identical with the Definition or supposition which is first laid down. Hence Cotes Theorem is the Same with the Definition of a Circle. Credat Judaeus Apella There is no real Truth but relative to | the Mind and to Suppositions arbitrarily made by it. Hence all Mathematical truth is purely Speculative. Yet he acknowledges it is of Use when properly applyed to Physics. But that Astronomy Optics & Mechanics are the onely parts of Physics which admit of such application.

Physical Truth is real & has no dependance upon the Mind particular Effects are by induction referred to more general Effects which we call Causes but in reality we are quite ignorant of Causes and cannot so much as form an Idea of a Cause these General

Effects are what we call Laws of Nature and we cannot go a step beyond them.

Discourse 2d The Theorie of the Earth to pag 168. The Structure of the Earth is altogether unknown to us except the Surface and a very little way below it. Even a great part of its Surface remains undiscovered. We observe a great part of it covered by the Sea whose waters are continually agitated being constantly carried from East to West by the tydes & in various directions by Winds and Currents

The bottom of the Sea appears to consist of the same Materials as the dry land, & to have the like unevennesses. What are Islands but the tops of high Mountains whose bases are under Water. The Strata of the Earth are commonly horizontal but we find sometimes heavier matter above & lighter below The whole Surface of the Earth is covered with Mould which seems to be the remains of Animal & Vegetable bodies corrupted & divided into so small parts as that their organization is imperceptible. The Strata that ly deeper are filled with Marine Substances shells Bones fishes &c.

The greatest ranges of | Mountains are towards the Equator where by reason of the diurnal Motion of the Earth the agitation of the waters is greatest. In tracts of Mountains we observe that the Salient Angles of one correspond to the reentring Angles of another Contiguous to it.

The great Tracts of Mountains in the Old World run from east to West in the new from North to south.

There are in many places even very high Mountains such vast collections of Marine Substances that they cannot be supposed to be gathered together but in a long Course of years or perhaps Ages. From these Phenomena it seems probable that the Waters of the Sea have formed the Mountains & have once covered them & gradually left them. that currents have formed the Valeys.

The rest of this Vol contains Proofs of the Theorie of the Earth in 19 Articles

Art 1. Of the Formation of the Planetes

Art 2, 3, 4, 5. The Systems of Whiston Burnet Woodward & some other Systems viz A Memoir of M Bourguet printed Amsterdam 1729 with his Philosophical Letters, to him we owe the above Observation of the Angles of Mountains which is a key to the theorie of the Earth. Leibnitz Protogea Act Leips 1683. Scheutzer Memoir Acc Reg 1708 Stenon Dissert de Solido intra Solidum. Ray

Art 6 Geographie Art 7 Of the Production of the Strata

Art 8 Of Shells & Marine Substances found in the Earth. A Memoir of Reamur presented to the Accad Royal 1720 wherein he Mentions a Mass of Sea Shells in Tourain 25 leages from the Sea of above 130,000 000 of cubical Toises |

Art 9 Of the inequalitys of the Surface of the Earth 10 Of Rivers 11 Seas & Lakes 12 Of the Tides 13 Of the Inequalities in the bottom of the Sea and of Currents. 14 Of Stated Winds 15 of irregular Winds Hurricanes 16 Volcanos & Earthquakes 17 New Isles Caverns & Perpendicular fissures. 18 The Effect of Rains, Marishes, of Subterranean wood & Waters. 19 of Changes of Dry land into Sea & Sea into Dry land

Vol 2 Of the Generation of Animals

There is no certain Criterion (according to this Author) by which an Animal is distinguished from a Vegetable there seemes not onely to be no Chasm between these two kingdoms Nor even room for a line to distinguish them.

The most remarkable Property wherein they differ from unorganized Matter is the faculty of Multiplying themselves by continual Generation of new organised Bodies like those which generate them.

We can conceive a Mould by which as many bodies as you please may be formed like to a given one in their external form or Surface. Tho it is not so easy to conceive a Mould for the interior form of a body yet Buffon thinks the difficulty of conceiving this is entirely owing to the limitation of our Senses which present onely the Exterior of bodies We must acknowledge that some Qualities of Bodies reach the inward Substance as Gravity. And if our Senses could perceive this property of bodies as well as their External form we might have as little difficulty in conceiving how that internal form may be given to bodies which makes them gravitate as we presently have in conceiving them figured into Cubes or Cylinders.

Some Bodies seem to have the power of Assimilating others to themselves As fire which turns every Combustible thing into fire

[3r] (we may add fermenting & putrifying bodies) An organised | body has this Assimilating power & turns its food into its own Nature; By drawing into its inward Substance forreign Matter which expands all its parts proportionally without altering its form or organization. And if we suppose some of the parts of such an Organized body to be similar to the whole these will when properly nourished produce an Organized body of the same kind. And if all the parts which nourish an Organized body be similar to the whole then every part of it may produce after its kind as it is in the polypus and in some trees & vegetables.

When Animal or Vegetable Bodies have more food than is sufficient for their Growth it may be presumed that among the Superfluous parts excreted there will be numbers of these small organised parts which are similar to the Whole and which of Consequence may be proper to produce Animals or Vegetables of the same kind.

In the food we take into the Stomach there are a great many organized parts as our food is mostly vegetable or animal. A great part of what is unorganized or dead Matter is carried of by the common Excretions. The chile contains most of the Organized Molecules of the food which mixing with the blood is in the course of Circulation farther purged of its inanimate & unorganized parts by the several Secretions of Perspiration Sweat urine &c the Organised Molecules being carried to all parts of the body along with the blood every part draws into itself what corresponds with ⟨it⟩ and by that means grows, not by an apposition of these molecules but by an intrasusception of the Whole into the whole so that these Molecules pervade the very Substance of the parts that they correspond with as the force of Gravity does and enlarge them in

[3v] all their Dimensions without altering their form | When the several parts of an Animal Body are thus extended to near their full dimensions they become more rigid and less fit for intrasusception of more Similar parts but the blood still conveys such Molecules to all parts of the Body, and the Superfluous ones are rejected to some receptacle destined by Nature for that Purpose as the Testes seem to be in Men and the Overea in Women. It is therefore to be expected that the Liquors thus secreted from the blood should contain Molecules fit for composing an Animal Body which Unite together by the same Power by which they enter into the parts

that correspond with them in the Nutrition & growth of Animals. In some Animals there is no distinction of Sex no Copulation Necessary. A single Animal produces its kind but in the Human Species there is a Seminal liquor in both Sexes which must be mixed & produces a Male or female according as the Male or female Seminal liquor prevails. As these Seminal liquors contain the Molecules proper for making the several Parts of the Animal. They by a natural Power or Virtue make various Conjunctions which seem to have life and Motion & perhaps really have so. these may be considered as the outlines of an animal which the Seminal Liquor is still sketching hence the disposition of all Seminal Liquors to produce Animalcula

Chap 5 The Systems concerning Generation Explained The Chief of the Modern ones are that of the Ova advanced by Harvey Supported by Malpighi Valisniere DeGraaf &c & that of the Animalcula in Semine Masculene by Lewenhock

II

Experiments Upon Seminal Liquors

The Sum of these is that in the Male Seminal Liquors of Men there were found long filaments that seemed to be holow transparent tubes and to have something moving within them in some hours as the liquor grew thiner from these tubes there issued out globular or oval animalcules tied to the tubes by a slender filament. which had an undulating or vibratory Motion. by degrees they were disengaged from the tubes and drew after them the filament which had then some appearance of a tail. They at last get free of this tail, and then their Motion is quicker and their size less the tail he thinks is no part of the Animalcule. Some of these Animals were round others oval some oblong & thick at both end this last kind sometimes divide in two which continue their Motion Still. Much the Same Phænomena were observed in the Male Seed of Dogs Rabbets Bulls &c

The Female Seminal Liquor is contained in certain glandulory bodies which when the female is fit for Generation appear upon the Ovaries of females & gradually Swell & become reddish and at last open & distill a wheyish liquor into the Tubæ fallopeanæ which when the female is hot grasp the Ovaries. These Glandulous bodies are quite distinct from the Ova as they are called & when they have distilled their liquor they shrink and disappear leaving a kind of cicatrice where they were. When these glandulous Bodies are ripe the liquor contained in them is filled with Animalcula precisely like those in the male Semen. The Testicles and Ovaries Steept in Water in a Sealed bottle produce Animals of the same kind. So does roast veal Pepper & many other Seeds of vegetables and parts of the flesh of Animals These Animalcules are imperfect Animals they do not generate nor produce their kind it is uncertain whether they have any Sensation, they are Organized bodies. which serve for the Nourishment or Composition or Generation of perfect Animals. The Seed of a Plant is not a Plant. The Egg of a foul is something intermediate betwixt an Animal & vegetable so probably are these Animalcules See Needham Microsc Obs 1745. 1747 |

I have seen (says M Bouffon) in the Ditches of the Count de Maurepas his Castle of Pontchartrain, carps well attested to be 150 years old, as agile & lively as ordinary Carps

All Quadrupedes Covered with hair are viviparous And all those covered with Scales are Oviparous. Vipers can hardly be considered as an Exception to this rule since they form Eggs but retain them in their Bellys till the Young be excluded. The same may be said of Salamanders. Mem de l'Accad Anno 1727

Query. Are not the Elephant the Rinoceros, the Whale Dolphin & all the Cetaceous Kind Exceptions to this Rule

Among the Animals that Copulate some seem onely to touch as the Cock the Sparrow the Pidgeon, altho' other Birds truly copulate as Ostriches Ducks Geese and have a Member of considerable Size for that purpose. Male fishes sometimes turn upon their Back to meet the Belly of the female, but it is not the love of the female that attracts them but of her Spawn So that there are some Animals that have Sex and parts proper for Copulation others that have Sex but no parts proper for Copulation There are some Animals that are both Male & female & Copulate as Snails some that gender without Copulation & yet do copulate as the Pucerons. Bees & Ants Male Female & neutral

All Animals in order to produce their Kind form Some intermediate Productions. in Viviparous females the Growth & repletion of the Seminal Glands their ripening opening and distilling their Contents into the horns of the Womb which at that time do embrace

Part One: *Natural History* 89

them. the forming of the Placenta & its Union with the Womb the Umbilical Chord. Are intermediate Productions in Males the formation of the Semen & the Animalcula in it. the Inflation and Erection of the Male Organ. In Oviparous Animals the Formation of the
5 Egg. In insects that undergo transformation there are more intermediate productions the Egg produces the Caterpillar. The Catterpillar produces the Aurelia. the Aurelia produces the Butterfly. The last production is intended in all the preceeding and they are so many Steps towards it. Of these intermediate Productions some are real
10 Animals as the Caterpillar Some seem to be something betwixt animal & Vegetable as the Egg The Placenta &c

Almost every Animal except Man has a fixed Season for Generation

The Loves of Birds are commonly in the Spring. Fishes Shed
15 their Spawn in the hot Months. Cats in Jan. May & September. Wolves & foxes in Jan. Deer in Sept & October

Every Species has its time of Gestation. Horses & Asses from eleven to 12 Months Women Cows & Hinds nine Months. Foxes & Wolves five months

20 A Hen will continue to lay prolific Eggs for three weeks or a Month after her Separation from the Cock |

[2r] Dogs nine weeks Cats six weeks Rabbits 31 days

The Decomposition of Animal & Vegetable Bodies produces many animalcules and vegetables which seem to be imperfect & to
25 have no power to continue their kind. The Milt of the Calmar produces a kind of live Pump which moves of itself. Barley Water produces A kind of eel like animalcules which while kept in Water appear to move but when dry seem dead and may be made thus to dy & live alternately for years together. For this purpose we
30 need onely steep the Barley 10 or 12 hours in water & separate it into its filaments

Paste made of Flour engenders little Eels which seem onely to be sacks full of other little eels as appears upon opening them with a lancit

35 The filaments in the male human Semen seem to be vegetables which in a little time produce innumerable little Animals

The Womb in Females seems to make a particular System by itself but whether of an animal or Vegetable Nature is difficult to say. This System lyes dormant & inactive like a vegetable Seed till it is excited by copulation. Which is to the Womb what moisture & heat to the seed of a Plant. The Concurrence of the Male & Female Semen gives the Stimulus, immediately the Womb begins to grow, & enlarge in all its dimentions. The Menses or even the Lochia from a preceding birth cease. Whether the first Elements of the Fœtus are from the male or female Seed or both or whether it grows out of the womb it self once excited by these is a mystery not yet cleared, The modern Theories suppose the Womb and Fetus to be at first unconnected and that the Ovum or Animalcul in the course of its Growth shoots forth vessels which unite with those of the Uterus and coalesce into one body as when a fly darts its Egg into the blade of a vegetable the egg or liquor conveyed along with it form an Uterus which comes to be connected with the Egg to nourish it and to grow as it grows. both in thickness and capacity. But for any thing we certainly know the first rudiments of the fœtus may grow out of the Womb by the Placenta as the pea does out of the Pod or the Plantule out of the parenchyma of the Seed. However this be the Womb & Fœtus either are from the beginning or soon become parts of the Same System, as much as the Stomach and the Man. or the tree & its Root. or the fruit & the branch By the Medium of the Womb the child and Mother are united. Whether a Magot bred in the leaf of a plant is united to the Plant or onely lodges in it is not certain. Is not the Aurelia the Womb of the Caterpillar & the butterfly the Fœtus of that Womb. In this case the Womb survives | the Mother and becomes itself a kind of Animal, which gives nourishment & life to the Fœtus. Perhaps in other Generations the womb is first enlivened and then the Fœtus.

Vegetable Life seems to preceede Animal Life and to be a necessary step to it. A Fœtus is at first rather a Vegetable than an Animal In the Semen of Animals and other fluids that produce Animalcules there is first the appearance of branches or tubes from the knots & swellings of Which Animals do spring forth. The Tænia Ascarides and other worms bred in animal Bodies seem to be formed by the active organized parts which abound most in the Semen but are likewise found in the Stomach liver & other parts. When the parts of Animal or Vegetable Bodies are in their Natural State the Organized parts are fixed & entangled by the unorganized & so shew no signs of Life or motion but when

Part One: *Natural History* 91

these parts are resolved and decompounded by being steept in Water the Salts are dissolved the Oyls separated and the organized parts begin to shew their Activity the more they are disentangled the less they are in size and the more active & this activity may become so great as to render them poysonous. The liquor that comes from the teeth of a Viper seems to owe its poysonous nature to these organized parts it contains which are extreamly active and perhaps the same may be said of the poyson of a Mad dog.

M Parsons Human Physiognomony explained

Buffon allows to animals Sensation and appetite. The Sensations first act upon the Nerves and by their means upon the brain which is the organ of internal Sensation & Motion. The motions of the Animal are performed by the reaction of the brain upon the Nerves. Which retains impressions made upon it and reacts not always instantly when it is acted upon by sensation (like inanimate Matter) but by laws peculiar to it. Brutes therefore have no memory no understanding no Ideas. Ideas are formed by comparing our Sensations and reasoning by comparing our Ideas. Brutes retaining past impressions and receiving present ones are acted upon mechanically by the joynt force of the impressions that are made upon their brain they retain past impressions but they do not distinguish them from the present and consequently may be said to have a material reminiscence but have no Idea of the past and therefore no memory. A dog sees meat in his masters hand which stirrs his appetite and prompts him to jump at it, but the impression of pain felt for former misdemeanours of this kind remains upon his brain & raises his aversion | these two impressions of desire and fear the one recent the other of an older date but retained in the internal Sense draw contrary ways and the animal necessarily moves or is at rest according as the restraining or the impelling force is greater

The action of external objects is received and modified by the particular Senses lodged in the nerves. This action so modified is carried to the brain where it is farther modified and retained. And according as it is relative to any of our Wants or appetites or aversions it is followed by a reaction of the brain upon the nerves which produces motion. The internal Sense differs then from the external first in its capacity of retaining impressions secondly in its reacting upon the nerves. it desires craves and makes an effort to obtain what it desires The Eye hath so large a Nerve that it may be considered as a production of the brain and it has more likeness to the internal Sense than any other it not onely receives impressions or objects but it reflects them & expresses desire appetite and other cravings. The other Senses are almost entirely passive, capable of receiving impressions but not of reflecting them

The Nerves that administer to Sensation, the brain & Spinal Marrow, and the nerves that administer to voluntary Motion. Make one System and are the Several parts of one Animal Machine. The first are the parts to which the power is applied, the second the fulcrum or fixed point, the third that by which the weight is moved or the effect produced. But the Brain is not onely capable of reacting as it is acted upon. It is it self active because it retains long the | Impressions made upon it by external objects by means of the Senses. If we should suppose the eye capable of retaining the impressions or images made upon it after the objects are removed it would then see things past as well as present. and be affected with desire or aversion not by these that are present onely but by the joynt force of the present and past. cooperating or by the difference of these forces where they act contrary ways and all this without memory. & in a purely mechanical way.

The Sense most relative to Knowledge is Touch which man has in a Superior degree to all animals. That which is most relative to appetite and instinct is smell which brutes have in a Superior degree to man.

Sight can serve for the purpose of Knowledge, but onely by means of the touch which corrects its errors. Hearing in Man is of the utmost importance as it is the instrument of Conversation & so becomes an active Sense

Brutes therefore have every faculty that we have except thought and reflexion they have feeling even in a higher degree than we they have the conscience of their actual Existence but not of their past Existence. They have Sensations but they want the faculty of comparing them that is to say the power which produces Ideas; for Ideas are onely Sensations compared, or rather associations of Sensations Every gentle Impression (Effleurement) upon the Senses is pleasure, every rude and violent one is pain. It cannot be doubted but that animals have more pleasure than pain for every thing that is aggreable to nature and tends to preserve it is pleasure.

III

2131/3/ II/16, 1r

October Read Contemplation de la Nature par C. Bonnet Amsterdam 1769. 2 Vols 8°

In a very long preface the Author gives a Sketch of two former works of his own. to wit *Considerations sur les Corps Organises* and *Essai Analytique sur les Facultes de l'Ame*. His Sentiments on these Subjects correspond very much with Dr Hartleys. His *Embranglements* with Hartleys Vibrations Memory & Imagination differ onely in degree from Perception. His System, that all the Operations of Mind take their Rise from certain motions of the Nerves, leads him into fatalism, as it did Hartley. Because his System might bring him under the suspicion of Materialism, he declares his Sentiments against it & says that he has given his Reasons against it in the Preface to his Essai.

He thinks every Organized Body must be the Work of God. That all have existed & been animated from the Beginning. That they pass through different states in which they have very different Organizations & different Powers depending on the Number & perfection of their Organs. Whether all the Organized bodies which descend from a common Stock by Generation were all originally cased up one within another in the first Parent or whether disseminated he does not determine, but inclines to the first Opinion, and is positive in Generation by Male & female, the organised Germ was contained in the Ovaries of the Female before Conception.

These Germs in the Ovaries of Virgin Females have a Circulation & Growth but can onely grow to a certain pitch, the resistance of the parts hindring their farther Growth as it does in old Men. Their impregnation gives an Irritation & Stimulus to the parts which enables them to overcome this Resistance. The Semen Masculinum not onely gives this irritation, but affords nutriment to the Germ, whence it is that children take after the father as well as the Mother. Monsters are produced Sometimes by the lesion of a Germ before or after conception, sometimes by the coalescence of two or more Germs. Any hereditary Monstrosity must be accounted for in the same manner as the likeness of the Children to their parents. |

[1v] The Universe is the Work of God contrived with the most perfect Wisdom and Goodness. It is one Whole All the Parts Simultaneous

or Successive, superior or Subordinate have a Relation to each other & to the Whole. Like a great Machine perfectly well contrived and executed, no one part however minute and insignificant could be wanted or displaced or be any other than what it is without impairing the perfection of the Whole. The various Parts of this great whole have different degrees of perfection, & rise one above another by insensible gradations so as to leave no chasm, no interval between the lower & the next higher It makes one Chain of which the Links joyn together. The least Interval would disunite the whole. If there were some parts of this great Whole endowed with Sense and Perception others altogether devoid of Both there must be a great Chasm. As the Author is unwilling to admit of this he is inclined to think that all Beings have some degree of Sense & perception.

The Universe he makes to be an essential Emanation from the first Cause, but thinks it is not eternal. There is nothing in the Universe of absolute evil, Nothing but what is the Cause or the Effect of some Good which could not have existed if that had not existed which we call evil.

The Author speaks of it as a point fixed that Venus has a Satellite.

Venus he says has Mountains higher & more numerous than those of the Moon

In speaking of perfection in general, he conceives every perfection as relative to an end the more excellent the end the greater perfection there is in the Mean. Query Does not this suppose a Perfection in Ends as well as what he calls relative perfection

The greatest corporeal Perfection is Organization. The highest degree of Spiritual Perfection is that of Abstraction or generalizing ones Ideas

The Ostrich at the Cape & in other temperate climates sits upon her eggs in the night and fetches food to her young till they are three or four days old when they are able to shift for themselves.

Because some Animals are almost fluid, or have no greater degree of hardness than a Gally he concludes that there is a fluid in the Nerves which is perhaps as subtle and active as fire or Light.

The Passions & moral Sense arise from some Reaction of the Soul upon the nervous System.

The Author distinguishes Sensation from Perception, but he thinks they have both the same Source & differ onely in degree.

Thus a moderate degree of Light gives us the Perception of Objects a greater degree dazles the Eye and gives an uneasy Sensation.

Is not the Extension of Matter, as well as its colour a mere Phenomenon relative to our Senses? Must not Matter be made up as all compound beings are of Unites, that is of simple Substances that are active?

The Yolk of an Egg is the Intestine of the Pullet, which at first is without its Body but as the pullet grows is drawn in. Hence the Author infers that the germ of the future pullet was in the Egg before fecundation & that what it derives from the Cock serves onely to irritate the fibres and produce their enlargement.

The blood in insects is a colourless fluid, which though it has nothing in it inflammable yet in many insects resists a degree of cold beyond that of our severest winters. |

[2r] Some insects as the Vine fretter are oviparous in one season & viviparous in another. About the middle of harvest they begin to lay eggs. It is then onely that the males appear, who are very small compared with the females, but very ardent in their Amours. The females gender without any intercourse with the male, through the Summer. Some of both sexes have wings & some of both want.

In the Mud of fresh Water lakes lives an Insect which on account of its likeness to the treble String of a Viol the Author calls Chanterelle. It is transparent from one End to the Other. It has a head & Mouth at one end & an Anus at the other Two vessels may be discerned that run through its whole length. One from its alternate dilatations and Contractions seems to be the Artery the other the Intestinal canal. If the Animal is cut into two, an Anus is in a few days formed in the upper Section & a head in the lower. If cut into 16 or any Number of pieces every piece becomes a Chanterelle with a head & tail & in a few weeks is as long as that from which it was cut. Another Species of the Chanterelle bears to be cut into two. But if cut into three or more parts the intermediate parts instead of a head and a tail get two tails & so perish for want of a head

The *Polype a Bouquet* is found in the Bottom of small Rills adhering to some fragment of a Vegetable. To the naked Eye it may be taken for small spots of mould in the putrefying substance. To the Eye aided by a microscope it seems to be a Cluster of little flowers in Bells. On more accurate attention there appears within each Bell an Agitation of the Water like that of a Miln, by which small Insects are drawn into the Bell & devoured by the Polype. Sometimes you see a Bell detach itself from the cluster, swim away and attatch itself to another place where it prolongs its stalk and carries on its operations A Bell sometimes closes into the form of a button. This button in a short time splits into two buttons lengthways one beside another, which two buttons open into bells & are perfect polypes. There is another way in which this animal propagates. Upon a Stalk or branch of the Polype a small bulb appears which increases in its magnitude very fast like a Gall nut upon a Plant. At last it is detatched from the Stalk, and attaches itself to some other body by a small Stalk which lengthens very fast. This bulb or button divides in two by its length, each of these divide into two again, this division & subdivision goes on, at last the Smaller buttons divide no more but expand like a Bell the larger continue to subdivide till they are brought down to the proper size by this means a Cluster consisting of 20 Polypes will in 24 hours divide into 100

Another Species of Polype is in the shape of a funnel Polype d'Entonnoir: and multiplies in a different manner. A transverse belt and oblique appears upon the funnel. The part above this belt separates from the part below and forms a new polype

The Polype *a Bras* multiplies by branches growing out of it; it multiplies by Cuttings. Cut it into as many parts as you please each becomes a polype. If you cut off its head, you may put it on and it joyns again as if it had never been cut. You may put the head of one to the body of another, & make up a polype of the parts of others put together *undique collatis membris* |

[2v] In the Animals we are wont to account more perfect, the Ovaries are situated in a particular part of the female. These are a cluster of eggs and every egg is the Germ of a future animal of the Species. In the Polype a bras the whole body seems to be an Ovary, and the eggs need nothing to their fecundation but to be placed in certain favourable circumstances, which they attain perhaps in ways we are ignorant of, and among others by being separated by cutting

The *mill pie a dard* has at first onely six feet; it grows in length, gets more rings and more feet and at its full growth has 200 but undergoes no Metamorphosis.

The Author to account for the wonderfull operations of Bees, Beavers &c hazards a Conjecture that though they have no innate Ideas they may have in their Structure certain Fibres which at the proper Season Suggest certain Ideas to their minds of the works they are to produce and the means of producing them.

The Author seems to have very just Notions of the Instinct of Animals and advances solid Reasons to shew that they are not endowed with the human faculties.

He mentions in a Note his having at the Age of 17 Anno 1737 begun a Correspondence with M Reaumur the French Pliny, & continued it for 19 years till that great Man died. He communicated to M Reamur all his discoveries in Natural History, in which he was very assiduous. His letters to Reaumur which would have made a great Volume were taken out of Reaumur's Cabinet after his Death and never could be recovered by him. Many Discoveries in Natural History contained in these letters have since been claimed by Authors who perhaps knew nothing of their having been first discovered by him. This he says he does not regret.

This Book is full of those Sentiments of pious Admiration of the Wisdom of God which the Subject naturally inspires. It is wrote in an elegant & entertaining manner. The Author has many Strictures against Buffons organised Molecules.

Read of the Same Author Paligenesie Philosophique 4 Vols 12°. This Work begins with an Ample Account of the thre⟨e⟩ former Works. & is intended as a supplement to them. The Doctrine mainly inculcated in this Work is that the Souls of Brute Animals, are immortal and pass into a future State at their Death, in which their faculties are enlarged or rather unfolded. The Author mentions his *Insectologie* published Anno 1745 a Paris As containing a great many of his Observations on Insects. Another Work quoted by him of an Anonymous Author is Essai de Psy⟨c⟩hologie, ou Considerations sur les Operations de l'Ame London. 1755. A third is of the Abbe Spallanzani Professor of Philosophy at Modena, & Member of the Royal Society at London Prodromo di un Opera da imprimersi sopra le reproduzioni animali 1768. from which the Author takes the Facts concerning the Snail and concerning the water Salamander, after mentioned.

The Garden Snail called by the French Escargot, has a brain

divided into two lobes, from which several Nerves and among others optick Nerves proceed

The Eyes are in the points of the horns and consist of Coats three Humors and an Optick Nerve as the human Eye does. If one or more of the horns are cut off they are reproduced in all their parts so perfectly that the nicest Anatomist | can perceive no difference between the reproduced Horns and those which never were Cut.

Nay if the whole Head of the Snail is cut off a new head grows out of the body as compleat in all its parts as the Old. Some Authors have expressed their disbelief of this. The Author has no doubt of it. He quotes here a Supplement to Bomarus his Dictionnaire d'Histoire Naturelle.

The Water Sallamander is a Quadruped. It has bones covered with Flesh The Spine and Tail are made up of Vertebræ as in other Quadrupedes. The Animal has some resemblance to a Lizard and is amphibious. In its four Leggs there are 99 bones articulated as in other Quadrupeds, with Marrow Periosteum ligaments Muscles Nerves Arteries & Veins Skin & Epidermis &c

If the tail or any number of its vertebræ be cut off they are renewed if the four legs are cut off other four grow out of the body perfectly like the former. The Abbe assures us that he has seen six successive reproductions of this kind in the Same Animal. In a Salamander full grown the reproduction is perfected in a but a year but in young ones in a few days

Mr Herissant de l'Accademie Royal de Sciences is quoted for a very curious Dissertation on the Composition and Growth of Bones, Shells, Corals &c Mem de l'Academie 1763. intitled Memoires sur l'Ossification

Another work of the Author quoted is Recherches sur l'Usage de feuilles dans les plantes.

PART TWO
Physiology

I

2131/7/ n/19, 1r

1 There is in Animals an Active Principle that has the Power of beginning Motion. All the Endeavours of Physiologists to Account for the Motion of the Heart & Arteries in the Circulation of the Blood of the Intercostal Muscles & Diaphragm in Respiration appear upon Examination to be utterly insufficient, and indeed to be puerile & unphilosophical. The passions of Shame fear Anger excite various motions in our bodies, the Appetit of Hunger & Lust nay the very Ideas of Certain Objects produce motions peculiar to them. Whatever is the occasion of these various Motions the cause of them must be something that has the power of beginning of Motion Sweet Oblivion winding round his head

II

Of the Involuntary motions of Animals

2131/7/ n/2, 1r

That the involuntary Motions of the Muscles are owing to some Irritation occasioned by a Stimulus of one kind or another, seems to be the most probable Hypothesis yet advanced on this Subject & I think the onely one that deserves examination.

I think it will be allowed that most of the Muscles of involuntary Motion may be excited to action by Some Stimulus

Thus I have known a piece of Tobacco taken into the Mouth immediately excite a Motion to Stool. It seems to be chiefly irritation that Catharticks and diureticks and Sternutatives do work as to the last I think there can be no doubt. Friction and other Irritation occasion manifest Exertions of other Muscles. Is it not probable that all these Medicines which are called hot, do increase the heat of the Body by irritating the blood vessels and increasing the Strength and Velocity of the Circulation which does alwise increase the heat of the body, tho in a manner we cannot account for because in other cases we do not see that heat is raised by the Motion of fluids unless in fermentations and Solutions.

Coughing & Vomiting seem plainly to be provoked by some Stimulus. Inflammations seems to proceed from some Strong Stimulus in the part inflamed which increases to a great degree the pulsations of the Small Arteries there and by that means distends the part and if the irritation be violent it occasions a quicker Circulation of the whole Mass of blood the Irritation being communicate.

We see that severe friction in any part of the body has the Same Effect as a beginning Inflammation & if the friction were continued it would no doubt produce Inflammation at last

May not fevers often at least proceed from an Irritation either of the Blood Vessels or the Lungs. For any Stimulus is easily communicated from the one of them to the Other Whatever causes quick breathing so soon followed by a quick pulse and a quick pulse will soon Cause Stronger or quicker breathing |

From these Instances it seems to be a general Rule that those Involuntary Motions that are Occasional are commonly excited by some Stimulus and these which are constant are increased by the Same Means. It may hence be not improbably conjectured that the Constant involuntary Motions of the lungs Heart Artery and Intestines are owing to some constant Cause similar to that which increases them occasionally that is to Some Stimulus

To this May be added the Experiment related by some of the Hearts of Some Animals when taken out and their Motion quite Ceased, that upon pouring Warm water upon them they will recover it again and continue to perform their Systole and Diastole for Some time. May not the Warm blood it self have the Same Effect.

On the other hand it must be acknowledged that there are considerable difficulties attending this Hypothesis.

First it may be said that we can affix no other Idea to the Word Stimulus but that of Pricking or the Sensation arising from pricking. We do not indeed commonly call it a Stimulus where both these do not go together. Now the last of these does not accompany the Motions in Question. For we feel nothing of this kind in the Circulation of the Blood the Motion of the Lungs or Peristaltick Motion of the Gutts. And a Stimulus that is not felt or perceived Seems to border upon a Contradiction. Or at least to signifie onely an unknown somewhat. like an Occult Quality. Or will it be said that there is a Sentient Principle in us distinct from the Mind which feels this Stimulus and is affected by it. For I think it cannot with any propriety of Language be said that mere inert Matter is acted upon by a Stimulus. Or perhaps it may be said that tho' the Mind had in the beginning a perception of this Stimulus yet like other feelings of pleasure and pain it becomes insensible by habit

But this last hypothesis labour(s) under this prejudice that it seems a general rule that where a Stimulus by habit & use becomes insensible it loses its effect so those that Sneeze upon taking Snuff give over doing so when it comes to a habit

III

Of Muscular Motion in the human Body

I take it for granted that the voluntary Motions of the Body which are perceiveable by our Senses are all performed by the Contraction of Muscles. The Muscles are bundles of small Fibres, wrapped round in Coates of the cellular Membrane, which having their two ends fixed in different parts of the Body, & having the Power of contracting their length; by that Contraction bring the two Parts to which they are affixed nearer to each other, & by that means produce all the Motions of the Body which Nature has put in our Power.

To enumerate all the Motions produced by this Contraction of Muscles, would be to enumerate all the labours of Mankind. By this we sit and rise, and walk & run and danse. All the Labour of the hands is performed by the Contraction of muscular Fibres. The Plough, the Spade, & the Hammer, The Chissel, the Pen, & the Pencil, are managed by this simple Engine. In every Mechanical Art the Contraction of Muscles is the great Spring that gives Motion to the whole. By this Palaces are raised & Ships built in which we sail round the Globe by Sea, or over Mountains in Canals by Land. By this the Pyramids of Egypt, & the Wall of China were reared to be everlasting Monuments of human Power, & of human Folly.

What is intended in this Discourse is, Some general Observations first on the Mechanism of the Muscles by which they are adapted by their Contraction to produce all the Voluntary Motions of the Body; Secondly upon that Power, whatever it be, by which they are contracted, & Thirdly upon some of the Accidents of this unknown Power, particularly those by which | it is affected in the Decline of Life

When we consider the Muscles as mechanical Engines, intended to produce certain Motions, we may judg of their fittness for that end, without going beyond the Limits of human Understanding; because there is nothing better understood than the Laws of Motion, and the Effect of those Laws when applied to mechanical Engines.

The whole System of the Muscles, may, I think, be compared to the Tackle of a great Ship, by which all the Movements are made in Masts & Yards & Sails & Helm, which are necessary in the Course of a Voyage. The Variety of those Movements is very great, & is known onely by the expert Sailor, who has spent his Life in acquiring this Knowledge. For they must be adapted, to every Variation in the Direction and Strength of the Wind, to Tydes and Currents, to Shoals & Rocks & Promontorys that ly in the Ships way. All those Movements are performed by the machinery of Ropes and Pullies simple or compound; so ingeniously contrived, and adjusted in their Power and Direction, as to produce the Motion intended in the easiest and most expedite Manner.

A great Ship of War is, I think the Noblest Machine that human Invention can boast. It is not indeed the Invention of one Man nor could possibly be; but of the combined wit & Experience of thousands in a Succession of thousands of years; by which, from the Bark or hollowed Trunk of a Tree, it has grown up to the perfection in which we now see it.

The Motions of the human Body in the various Exercises and Employments of Mankind are more in Number, & more various in their Nature, than the Motions of the ship in a Voyage They are performed in much less space, within the Compass of the Body; and all by the Machinery of Muscular Fibres, which are a kind of Ropes or Cords framed by Nature

The Number of the Muscles is very great, & their Form various corresponding to the Number & Variety of Motions to be performed.

Some Motions have great Extent, as those of the Arms & Legs, | others have onely a Small Extent, as that of the Vertebræ of the Back. In some, great Force is required, in others very little. In some Joynts, the Motion is onely forewards and backwards, like that of a door on its hinges, as in the knees and elbows. In others the Motion is in every Direction like that of a ball & Socket as in the shoulder and Thigh. Some Motions are complex, and require the Exertion of many Muscles acting either simultaneously, or successively in one certain Order, as in chewing & swallowing, in speaking & singing, and in many other operations.

That the Muscles, in a perfect human Body, are most accurately fitted for their different purposes, in their Matter & Form, in their Strength or Weakness, in their Origin & Insertion, in the length and various directions of their Tendons, according as the Laws of Motion and the Principles of Mechanism require, may be seen in Dissections, and has been Demonstrated by Anatomists.

It is no Wonder that Galen who had been educated in the Principles of the Epicurean Philosophy, when he came to be acquainted with what was then known of the Anatomy of the human Body, saw such manifest marks of Design and Wisdom in its Structure, as forced him to renounce his former Creed, and to acknowledge that there must be a Wise & intelligent Author of Nature. Many & great Discoveries with regard to the Fabrick of the Body have been made since the time of Galen, and the Laws of Mechanism are much better understood. But these Discoveries & Improvements tend onely to give a greater display of the Wisdom of Nature, in the Fabrick of this Machine, without detecting any Defect or Superfluity in the Means employed for answering the Ends intended. Indeed, of all the Works of Nature that fall within our View, there is no | Part that exhibits, in so small a Compass, so many evident marks of wise and good Contrivance as the Fabrick of the human Body; & therefore, it deserves the Study of all who are delighted with the Contemplation of Nature, not merely as a branch of the medical Art, but as one of the most fruitfull Branches of natural Philosophy.

In every Muscle there is a part of its length which, in its inactive state, is soft and fleshy, but when the Action of the Muscle is exerted, this fleshy part, called the Belly of the Muscle, contracts in its length, swells in its other dimensions, and becomes hard and rigid. The whole active force of the Muscle is in this fleshy part, called its Belly, which in some makes the whole or nearly the whole length of the Muscle, in others but a small part, the rest being dead Rope called the Tendon.

The Tendons of the Muscles are Ropes of Natures work, & in their pliableness and smoothness far exceed any made by human Art.

The great Impediment to the perfection of Machinery wrought by Ropes is, that a great part of the force applied to them is consumed in overcoming their rigidity & Friction, even when this is in part remedied by their moving on pullies or rollers wherever they touch other Bodies or are bent in order to change their direction. The Tendons are in contact with other Bodies through their whole length & in every point of their Surface. They are sometimes bent in a more acute sometimes in a more obtuse angle as there is occasion to change their direction; yet, in these circumstances, they slide along without the use of pullies, and with no sensible friction or impediment. Sometimes the Tendons are bent by passing under Ligaments that surround the Bones, sometimes by passing over protuberances or through perforations in the bones that are made for the purpose

This Lubricity of the parts of a living Body by which the tendons slide with so little Impediment & in such Circumstances, seems to me to be one of the Wonders of our Frame. It seems to have some dependance upon the Principle of Life, for we see that Death immediately destroys it.

We may next observe the Wisdom of Nature in the position of the Belly or fleshy part of the Muscles. |

[5] This it is evident ought to be placed where it may have room to swell in the contraction of its length without Impediment & without hurting the contiguous parts by its pressure. This Intention is most admirably observed through the whole System of the Muscles.

It is most obvious in the hands and feet the legs and arms the most moveable parts of the Body; which having many joynts must have many Muscles. If the Belly of a Muscle were placed upon a joynt, it would be hurt by the Motion of the joint & be an impediment to that Motion. The hands and feet are full of joints, & are often pressed by external bodies and therefore would be an improper situation for the bellys of Muscles. Therefore Nature has placed the bellys of those Muscles in the Calf of the leg & in the interstices of the joynts of the Arm & the Thigh, where their Swelling is neither impeded nor can hurt any of the contiguous parts.

It has been observed that the Action of the Muscles tends very much to promote the Circulation & Secretions of the Fluid parts of the Body. And indeed Experience informs us, that Active Employment is much more favourable to this Intention, than Inactivity & Rest. It is more to our present purpose to observe that the position of the fleshy part or Belly of the Muscles is so contrived by Nature as not onely is most convenient for their free Exertion, but so as to contribute greatly to the comliness and beauty of the human Form. In the exterior and Visible parts of the Body it is the fleshy part of the Muscles covered by the common Integuments, which makes the greatest part of the Contour or outline of the

Part Two: Physiology

Body. The beauty of its Form therefore must greatly depend upon the figure it takes from the bellys of the Muscles. The Ancient Statues of Venus of Apollo & of Hercules, in which all ages have admired the expression of Grace, of Dignity & of Strength, are [6] onely Copies of the Forms | which Nature, by the Disposition of the Muscles gives to a beautifull Woman, to an elegant young Man, & to a Man of gigantick Strength. Hence it becomes so necessary to the Statuary and the Painter,to know so much of this branch of Anatomy as to understand perfectly the appearance which 10 the various Muscles in the external part of the body, make to the Eye of a Spectator, both in their relaxed State, & when more or less contracted either by voluntary Motion or by passions of the Mind

It is the disposition of the Muscles in the human Countenance 15 that chiefly constitutes its beauty, and produces all that variety of Features by which the Passions and Temper of the Mind become in some degree visible

The Strength of the Muscles, or the Force with which they are able to contract their Length, is generally in proportion to the 20 quantity of Fibres of which they are composed and is extreamly different, but always proportioned to the work they have to perform. From the Figure of the human Body it necessarily follows, that the muscles must act at a small distance from the center of the Motion they produce. The Resistance to this Motion acts often at 25 a much greater distance. In such a case, from the Laws of Mechanism, the force of the Muscle must be greater than the Resistance to be overcome, in the same proportion as the distance of the Resistance from the Center of Motion is greater than the distance of the Muscle.

30 Thus, for instance, a Man of no gigantick Strength may, I suppose, support at arms length a weight of 30£. The Center of Motion in this case is the center of the Globular head of the shoulder bone. The Distance of the Weight from this center in an Arm of no extraordinary length may be 30 inches. The Distance 35 of the Muscle that supports the weight is not above three inches, that is, the tenth part of the distance of the weight. From this supposition it follows that in order to support a weight of 30£ [7] the Muscle must contract with a force | equal to the weight of 300£. A Man may support a very great weight if he can contrive

to place it so that it acts at a small distance from the center of Motion. Desaguliers who was a Man of no uncommon strength, standing upright upon a Table, by a Rope which passed between his feet through a hole in the Table, supported a Roller of 1900£ weight. The Rope being fixed to a harnessing that went round his Loyns

The Strength which in some cases is exerted by Muscles is indeed such as we would be apt to think sufficient to tear them in pieces considering of what soft & pliable materials they are made. But nature in every case has suited their Strength to the Stress they have to bear.

It is but rarely however and for a short time that we have occasion to exert the utmost degree of our power in muscular Motion The common Operations of Life require Art and Skill rather than great Strength, & we may overshoot the Mark we aim at by too great exertion, as well as fall short of it by too little.

In almost every operation we set about, there are certain muscles to be put in motion, and no others; they must be exerted in a certain order, and in no other order; & with a certain force & neither more nor less. But how do we acquire the skill to manage this machinery so Artfully? Would it not seem necessary that the person who performs such operations should before hand be acquainted with the whole System of the Muscles, that he may pick and chuse those onely that are fit for the present purpose, as a printer does his types, and that he may put them in that order and give them that degree of force which is required?

Suppose a Person to have a Fount of Types set before him with a printing press, & all the materials necessary for printing, & that he is desired to print such a Sentence: He sets about it and executes it with the Expedition & perfection of a Printer. Would it not be very wonderfull if we certainly knew that this person was perfectly ignorant of the art of printing and knew not so much as the figure of the letters? Would we not find ourselves under a necessity of concluding that this ignorant person was onely a blind Instrument in that Operation? That his hand was led by some invisible power that knew | the Art of printing? for that such an operation should be performed by mere Chance exceeds all the bounds of Credibility.

However Wonderfull the Phenomenon we have supposed may appear, we perhaps do not often consider that something no less

Part Two: Physiology

wonderfull may be perceived, in the most common operations of Men.

Thus an Infant sucks and swallows his Food with appetite when applied to the breast. In this operation thirty or fourty Muscles independent of one another are set to work These must work in a certain precise order, which is repeated in every draught. These Muscles are selected from the whole System of Muscles, whose Number is greater than that of the Types in the printers Fount. They are arranged in the precise order necessary for the end intended & in no other, and this repeated hundreds of times without Variation or Mistake.

We need not say that the Infant knows nothing of Muscles or their order or Contractions necessary in this Operation, any more than the ignorant Person was supposed to know of Types & printing Tools.

If we ask the Cause of the Infants sucking & swallowing so skillfully? Every one is ready to answer that it is Instinct. But what is this Instinct?

If it mean no more than that it is not knowled⟨g⟩e in the Infant. This is true, but we want to know what is the Cause not what it is not. There must be an Intelligent efficient Cause of a Work which shews a perfect knowledge of the internal Structure of the human Body as well as an ability to guide the infant blindfold to what is necessary to its preservation. We may therefore, I think, say without a Figure that this Instinct is the Inspiration of the Almighty. And the same may be said of every Operation which we ascribe to Instinct. It necessarily implyes the interposition ⟨of⟩ some invisible Power that leads us as it were by the hand in a way we know not.

Let us again suppose that this Infant after a dozen of years walks and runs, writes or plays on a musical Instrument. In all these Operations a Number of Muscles are set to work in a certain determinate Order How is this performed? It is not Instinct. It is Art grown into a Habit by learning and Use. But has he been taught to know what muscles are to be used in these Operations, and in what order? By no means. He is as ignorant of this as when he sucked the Breast. It appears therefore that Arts acquired by instruction and Exercise require the Direction of some invisible Intelligent Power no less than Instincts. |

[9]   Hitherto we have considered those Muscular Motions which are

guided by the Determination of the Will without any previous Act of the Understanding that we are conscious of: There are however many voluntary Motions in which some previous Perception of the Understanding is necessary to direct us to the Motion which the occasion requires, in order to prevent some sudden danger that leaves no time for Deliberation.

It is by some very accurate Perception of this kind that we are able to keep our Body so poised as to prevent its tumbling down by its own Weight.

It is a well known Principle in Mechanicks, that a Body stands upon its Base onely when the perpendicular from the Center of its Gravity falls within the Base, & that if it fall ever so little without the Base, the Body begins to tumble, & by a Motion very quickly accelerated falls to the Ground.

In all Exercises and Employments which require a standing posture, our two feet & the Space between them is the Base on which we stand, within which the perpendicular from the Center of Gravity of the whole Body & all that it carries must fall. And if by any Motion it verges ever so little beyond that Base, it must, instantly, by a contrary Motion be brought back to its place, or we fall to the Ground.

When a Man works with heavy Tools & with great Motion the Center of Gravity and consequently the perpendicular from it is continually shifting to one side or another; yet with all this violent Agitation, he keeps it within the small base on which he stands, & if at any time it verges ever so little beyond it, he has an instantaneous perception of his danger and by a Motion no less instantaneous, brings it back into its place. |

[10] This Power we have of perceiving the ballance of our Body is so like to our other external Senses, that it might very justly have been accounted a distin⟨c⟩t Sense, if it had been so much reflected upon as to require a Name.

In each of the external Senses, there is an Impression made upon the Body or on some part of it, which by our constitution produces a certain Sensation of the Mind, and that Sensation is by our Constitution accompanied with the Perception of something external.

In this Sense of our ballance, an Impression is made upon the Body by the power of Gravity, the moment that the Center of

Gravity goes beyond the Base, by this Impression an uneasy Sensation is raised, accompanied with the perception of our being in danger of falling in such a Direction. The uneasy sensation we feel and the perception of our Danger, are both the Effect of our Constitution; for if we were to wait till the inclination of our body in falling were discovered by any of our other Senses, the danger would be past remedy.

This sense of Ballance may be seen in a Child of two or three Months old. If sitting upon ones knee he begins to tumble, he immediately starts & endeavours to recover himself; But it is greatly improved by Use, in every Employment that requires its exercise; as in fencing, dancing, & in horsemanship; and still more in the useless Arts of Tumbling, Ballancing, and Ropedancing, which chiefly depend upon carrying this Sense to a wonderfull degree of acuteness by much practice. This sense of our Ballance is produced not onely by the impression made by the power of gravity but by any other Force which endangers the Ballance.

When a Man gallops round in a small Circle, in learning to ride, he has a centrifugal Force, tending from the Center of the Circle, & proportional to the Velocity of his Motion. This Force like a weight hung on that side, would through him from the horse if he sat upright, but he feels it the moment he is in danger, & by inclining his body towards the Center of the Circle, ballances this centrifugal Force. Nay, the Horse is | endowed by Nature with the same Sense. The Base on which he moves when he gallops is very narrow, perhaps not six inches broad, and the centrifugal Force would throw him without it, if he did not incline his Body the contrary way. The same thing may be observed in a Man who is expert in skaiting, who can turn with a very quick Motion upon a base that is not half an inch in breadth.

When we observe with what ease, and Grace those Motions are performed by those who are expert, and compare them with the Laws of Motion, we must be convinced that this Sense by which we perceive the least deviation of the Body from its Ballance, may by Use be brought to a degree of Accuracy which is hardly to be observed in any of our other Senses.

Nor is this Perception more acute, than the Exertion of the Muscles necessary to remedy the Deviation is quick & instantaneous. For this Purpose certain Muscles must be contracted, in a certain Order, & with a certain degree of Force. All this is done in the twinkling of an Eye, without any thought of Muscles or their order, or of the Center of Gravity & the Base to which it must be confined.

As the Wise Author of Nature hath given us this delicate Sense of our Ballance, & the power of those instantaneous Motions that are necessary to preserve it; so He hath framed our bodies with such variety of Flexures in all Directions, that on whatever side the Center of Gravity verges beyond its Limit, it may by a Flexure of the body to the contrary side be brought back to its place. Thus as the knees bend one way, the thighs the contrary, by making these contrary flexures at the same time, we ballance our body in sitting down & rising up and in many other Motions. When I stretch out my Arm with a weight in my hand; this weight on one side would destroy my ballance, if I did not at the same time | withdraw my Body to the opposite side, so far as is necessary to preserve the ballance. As the whole Trunk of the Body turns upon the Thighs, their joynts are made like a ball and Socket, by which the body can turn upon them in every direction so as to be a counterpoise to any Force that would destroy its ballance on whatever side it acts.

Although all voluntary Motion is performed by the Contraction of Muscles, we must not from that conclude that when no Motion is willed, the Muscles are inactive. *The Exertion of Muscles is no less necessary to rest than to Motion. In every position of the Body excepting perhaps that of lying prone* The reason of this is that there are so many Articulations in the Limbs, & in the Spine & Neck and these in a living Body have such Lubricity to facilitate their Motions that without the Exertion of Muscles, it would sink down to the ground like a Chain of many links. So we see a Man does if he is struck dead or deprived of all power of Muscular Motion in an instant, whether he be on horseback or on foot, sitting or standing.

Every Joynt has two contrary Motions, forwards & backwards If one of these Motions be made by one Muscle, the contrary must be made by an Antagonist Muscle, and to keep the Joynt steady without Motion in any position both these Muscles must act at the same time with forces so accurately adjusted as to be equipollent, that neither may be overcome by the other. That This requires a very accurate adjustment we may see by this simple Experiment. Let a Man put his two hands in the opposite Scales of a very nice ballance and try to press both with such Forces as to keep the ballance steady and without Motion. He will find this very difficult, if not beyond his power. Yet the same Man can keep a Joynt steady & without Motion for a long time, by a proper adjustment of the force of the antagonist Muscles.

The Case we have put, of two Antagonist Muscles by contrary forces keeping a Joynt without Motion, is the simplest that can be put. In Joynts that move in all directions like a ball and Socket, it is impossible that there should be a muscle to act in each | of these Directions, for they are infinite in number. But every Direction in which we intend to move the Joynt if it does not coincide with the Direction of any one Muscle, must at least be intermediate between the Directions of two or more muscles. In this Case two Muscles at least must join their forces to produce the intended Motion, and their Forces must be so adjusted to each other as that the direction of the Force compounded of two or more Forces shall be that which we intend. To keep such a Joynt without Motion in any particular Position a Number of Muscles and their Antagonists must act at the same time with Forces so accurately adjusted that the sum of the compounded forces in any one Direction may be equipollent to the sum in the contrary Direction. Like a Number of Ropes drawing at the same point in all different Directions, so exactly ballancing one another by contrary Forces that the point from which all of them tend remains perfectly at rest.

Thus we see that the human Body being made up of a number of Joynts & Articulations, could not rest in any particular Attitude if it were not supported by the constant Exertion of many Muscles. It is onely by a wonderfully accurate adjustment of the Forces of many Antagonist Muscles that it keeps its erect Form, or any other Position in which we chuse to continue without Motion. Nor is there any Position in which the Muscles destined for voluntary Motion are inactive, except that of lying prone, which therefore is the most proper for sleep or rest after fatigue.

The Tongue seems to be almost wholly composed of Muscular Fibres going in all the Directions of its length breadth and Thickness; so that we can make it long or short, broad or round, flat or hollow. It is wonderfull with what Volubility and Quickness it can by the contraction of its Fibres be suited to all the infinite Variations of Articulation & Voice, in Speech, yet so as to preserve | the peculiarities which distinguish individuals, so that every Man is known by his Voice as by his Features.

By the Lungs, the Aspera Arteria, the Larynx with its Epiglottis the Pharynx and Mouth, a Musical wind Instrument is formed by Nature in some persons far exceeding any of human Invention. It is by the Contraction of Muscular Fibres that this Organ is tuned to all the Tones Acute & Grave through several Octaves; and to all the Modulations Melancholy or Cheerfull, Solemn or Sprightly by which the Passions of the Mind are expressed

The Eye is an Organ contrived for the purpose of Vision, with consummate Wisdom; and to answer its end, is provided with various Muscles external and internal. By its external Muscles, it can be directed accurately to its object whether placed high or low, to the right or left. or straight before us. By these Muscles also the Axes of both Eyes are directed to the same object with perfect ease & great exactness It is necessary to distinct Vision, not onely that the Axes of both Eyes should be directed to the Object we look at, but that they should be kept steadily in that Position. This requires the cooperation and exact ballance of the antagonist Muscles.

I have known Persons, who either from an original Defect, or some hurt or Disease, could not command this ballance in looking at an Object directly before them. Their Eyes in that case had a quick vibration which hurt distinct Vision, and had a very disagreable appearance to a Spectator. Yet those Persons could keep their Eyes steady when directed to one Side, and therefore always turned the side of their head to an Object they looked at, and walked with the head to one side that they might see the Objects that were in their way. I believe this defect is uncommon, as, in the course of a long | Life, I remember onely to have observed it in two Persons. It shews however that in perfect Eyes there must be a very accurate ballance of the antagonist Muscles to keep them steady in every Direction.

There is a correspondence or Sympathy between the Muscles of the two Eyes, for which I think we can assign no physical cause. The muscles that move the two Eyes are as independent of one another & unconnected as any in the human body yet they always act in conjunction so as to keep the Axes of the two eyes nearly parallel. They are always both directed the same way, whether it be up or down, to the right or to the left. nor can we squint or give them different directions but by uneasy straining which some may learn by much practice. And it is observable, that this parallel motion of the Eyes takes place, not onely when performed by the Action of the similar Muscles, but when it must be performed by dissimilar or opposite Muscles. When we look up or down the similar muscles of the two Eyes called *lævatores* in the first Case and depressores in the second are the Muscles that act conjunctly; but when we look to the right or left. dissimilar or opposite muscles of the two Eyes joyn in their Action, the muscle called *internus* or *adductor* of one Eye joyning with the externus or abductor of the other Eye. We easily perceive that this Consent & Concurrence of independent Muscles is designed and effected by Nature because it is necessary to distinct Vision; but by what Means it is effected, is I apprehend beyond the reach of our Understanding. This Consent and Sympathy of different Muscles is not peculiar to the external and internal Muscles of the Eyes, it may be observed in the Muscles of the opposite sides of the face, which concur in their action in speaking, smiling, frowning, laughing, so as to preserve the similarity of the opposites sides of the face in all these Motions. Hence when any muscle on one side is disabled by palsy or otherwise, from concurring with its opposite, a real deformity in the Face is the Consequence.

Besides those Motions of the Eye by which it is directed to its object, it has other Motions which may be called internal, by which it is adapted to different degrees of Light, & different distances of the Object from the Eye. |

[16] It is adapted to the degree of Light by two antagonist Muscles one of which contracts the Pupil, & the other enlarges it, and both without altering its circular figure. When the pupil of a young Eye is observed in a very faint light, if then a very strong Light is brought before it, we may perceive it to be contracted to about half of its former Diameter. If the light be so strong as to be offensive after we contract the pupil as much as we are able by its own Muscles, we cover a part of it by the Eyelids.

In some Animals the Muscular structure by which the pupil is enlarged or contracted to adapt it to the degree of Light is quite different from what it is in Man. In Cats who are made to hunt their prey in the dark, the pupil when most dilated is circular, & almost as large as the Cornea. According to the degree of Light they contract its breadth, while its vertical diameter remains the same, so that in a strong light it is contracted to a hair breadth, & appears as a black line dividing the Cornea into two semicircles on the right & left. I think I have observed the same thing in Owls.

I apprehend that Cats can see their prey when there is no external Light at all, by a phosphorous Light emitted from the Center of the Retina of their own Eyes; which Light being refracted in passing through the Eye, falls in parallel Rays upon the spot to which the Eye is directed, and no where else. The Evidence of this is, That in the most perfect darkness you see the Eye of a Cat, when she looks at you but not otherwise, & the pupil of it in that case appears circular & in its largest dimension.

If the common belief, that a rattlesnake by fixing its Eyes upon a bird can charm it so as to bring it within its reach, I say, if this be a Fact sufficiently attested, it deserves, to be enquired, whether this Charm may not be performed by the birds perceiving some phosphorous Light like a Gloworm in the Eye of the Serpent which allures its approach to its own destruction

It is still disputed among Physiologists, by what muscular Structure the Eye is adapted to the distance of its object, so as to | produce distinct vision whether the object be near or distant.

It is evident that when the object is near the pencil of Rays which passes from every point of it to the pupil must diverge more than when it is distant; & therefore in order that these Rays be collected into a Focus at the Retina, they must either be more refracted in the Eye or the Retina must be farther removed from the Cornea by compressing the Globe of the Eye about its middle & so lengthening its Axis. To me the Opinion of Dr Jurin seemed very probable, to wit that the rays are more refracted when the Object is near, by a very small increase of the convexity of the Cornea, & that this is effected by the contraction of a small muscular ring which surrounds it & that the Elasticity of the Cornea serves as an Antagonist to this Muscle. That ingenious Author has shewn that almost the whole refraction of the Rays in Vision is made at their Entrance into the Cornea, & that a very small increase of its convexity is sufficient to produce distinct Vision of near Objects, and that no Motion of the Chrystalline is sufficient.

Not to enter upon disputable points; this little Organ of the Eye exhibits to us a very wonderfull Combination of muscular Motions. We no sooner will to look at an object, than one Set of Muscles is at work to direct both Eyes to it, so as their Axes shall meet accurately in the object, another set is at work, to adjust the aperture of the pupil to the degree of Light, and a third set, to adapt the eye to the distance of the Object. All this is done in the twinkling of an Eye, by those who think nothing, who know nothing of the Mechanism by which it is performed.

The Observations I have made are intended to shew, That the System of the Muscles, as a part of the Machine of the human Body, is contrived with perfect Skill & Wisdom for the End for which they are intended. They are the Instruments or Engines by which all our voluntary Motions are performed, in all the Employments, Arts and Exercises of Mankind; and they are perfectly fitted for this purpose; By the Matter of which they are made; By the Parts to which they are fixed both in their Origin and Insertion; By the various Contrivances of Nature for changing the Direction of the | Tendons, when necessary; By the most advantageous position of their Belly or contracting Part; By having their Strength adapted to the work they have to perform; By the Cooperation of many Muscles in complex Motions, either Simultaneous or Successive in a certain precise Order, a Cooperation which in some Cases is the Effect of Instinct, as in sucking & Deglutition; in other Cases is produced by Habit, as in writing, & in most Manual Arts; By their Instantaneous Exertions to recover the Ballance of the Body, the moment it is in danger of being lost; and By the wonderfully accurate Adjustment of the Force of Antagonist Muscles, in keeping the Body upright, & in keeping the whole or any part of it steady in any Position that is found convenient.

When we compare the Works of Men with the works of Nature, we may see in both sufficient Marks of the Wisdom, Skill & Contrivance of the Maker; But in this they differ, that in human Workmanship, we may see the whole Contrivance, & have as perfect and adequate an Idea of the Work as the Maker himself had. But in the Works of Nature, we see a Part & not the whole. There are Mysteries remaining which surpass our Understanding, as the works of Men surpass the Understanding of a Child.

Thus the Anatomist sees, that every Muscle of the human Body is most exactly fitted, by its Contraction, to produce the Motion intended. The structure of the Parts & the Laws of Motion demonstrate this. But if we wish to know how, and by what Power the Muscles are contracted in the very moment we will the Motion which their contraction produces. here we are at a Stand.

I think all that is known with respect to this is, That the Muscles are contracted by some Power communicated from the Brain by the Nerves, from which it is called *the Nervous Power* or *nervous Influence*. Every Muscle is served by some Nerve or branch of a Nerve whose fibrils are spread over all the muscular Fibres. And when the Communication between the Brain and any Muscle is intercepted by cutting the Nerve, or pressing it hard, we have no power to contract | that Muscle. I believe this is sufficiently confirmed by Experience. It implies, That some Power or Virtue is exerted at that part of the Brain or spinal Marrow where the Nerve has its origin, in that very moment when the Motion to be produced by it is willed, & in that measure and degree that is necessary to produce the Effect intended, That this Power is conveyed by the Nerve to the Muscle, and that by this Power the Muscle is more or less contracted.

These are Accidents of this Power which deserve to be known, as every thing does which has a Relation to it, but they leave us much in the dark. For we neither know what this Power is, nor how it is exerted, nor how it is conveyed by those Nerves onely which answer the present purpose, nor how it produces the Effect of contracting the Muscle.

These are Mysteries beyond the limits of our Understanding; and all the Attempts made to make them intelligible have been in vain.

The Ancient Theory of a Fluid flowing in the Nerves called *Animal Spirits*, and the more modern Theories of Vibrations in elastick fibres of the Nerves, or vibrations in an elastick Ether which pervades all Bodies, are all Hypotheses, and like other Hypotheses in Philosophy, labour under two defects. First They suppose the Existence of certain things of whose Existence we have no Evidence & Secondly, when they are supposed to exist, they do not account for the phenomena they are brought to explain. Neither the Exertion of this nervous Power at the moment it is wanted, nor its conveyance

Part Two: Physiology 119

by the Nerve, nor its Effect in contracting the Muscle, can be accounted for by any laws of Mechanism we know. It is something beyond Mechanism & of a Superior Nature.

There are other Powers of Nature, whose Cause and manner of Operation is beyond the reach of our Understanding, such as the Powers of Gravitation, of Magnetism Electricity & many more. Yet it is the business of the Philosopher to observe as far as he is able, every Circumstance relating to those unknown Powers, that may enable him to judge of the effects that they may be expected to produce in certain circumstances, or may enable him to apply them to usefull purposes in human life. |

[20] This we may do with regard to the Nervous Power, while at the same time we acknowledge that its Nature and manner of Operation are quite beyond our Conception.

First it may be observed of this Power that it is not exerted constantly, but occasionally & in consequence of the Will and Effort of the Person. I speak here onely of the nervous Power by which the Muscles of voluntary Motion are contracted. The Power by which the Muscles are contracted which produce the vital & involuntary Motions, such as the Pulsation of the Heart and peristaltick Motion of the Intestines seems to have no dependance on the will, but to have its alternate Exertions & Remissions, constantly succeeding each other, while Life continues. But the nervous Power, by which our voluntary Motions are produced, is dormant untill we will the Motion, and in that moment is exerted.

In this it differs remarkably from the Power of Gravitation & the other natural Powers of which Philosophy treats. They all act constantly, & equally in the same Circumstances without Dependance upon the Will of Man or of any Animal. I know onely of one exception to this, in what is said to have been lately discovered of the Electrical Eel, which has the Power of giving voluntarily something like an Electrical Shock to its Prey, or to any Animal it intends to hurt, not onely when any part of its Body is touched, but even at some Distance. This seems to resemble more the nervous Power than anything else I know in Nature; as it is dormant at other times, and exerted onely when the Eel wills to hurt its Enemy or to kill or stun its Prey.

What Relation the nervous Power has to the Will of the Agent is perhaps beyond our power to determine. All we know is, That in the sound state of the Body the Exertion of the nervous Power immediately follows our Volition. But whether the Mind, by its Volition, be the proper Efficient cause of that Exertion has been doubted. Malebranche and other Cartesians thought that it is not. and that the Mind, neither by its Volition nor by any other Operation can produce any Motion in Matter. But that by fixed Laws of Nature certain Motions of Matter are produced by unknown Causes upon Occasion of certain Volitions of the Mind They say therefore, | that the Mind is the occasional Cause of our voluntary Motions, but not the efficient Cause.

It may indeed be observed that when we will to produce any Motion, the Exertion of Nervous Power is not the thing we will. The external Motion is the onely thing we will. Yet Philosophers know that the contraction of Muscles is the Cause of the external Motion, & that the Exertion of nervous Power is the Cause of the Contraction of the Muscles. It is difficult to conceive the Mind to be the Subject of a Power, which is exerted not when we will that Exertion, but onely when we will something different which is onely the remote Effect of that Exertion. It is difficult to conceive our Mind to be possessed of a Power of which we know neither the Nature nor the Manner of its Operation.

We have a natural Conviction of our being the Cause of our voluntary actions, and therefore accountable for them. This Conviction, which is the Work of Nature, and of the greatest Importance in Life, ought not to yield to Physical or Metaphysical Speculations. nor indeed can the System of occasional Causes, though adopted, overturn it. For he that believes a certain Effect to be in his Power and exerts his Power to effect it, is undoubtedly in moral Estimation the Cause of that Effect and accountable for it. whether in Physical Consideration he be really the Efficient, or onely what the Cartesians call the occasional Cause. This Dispute therefore about occasional Causes, as it seems hardly capable of a certain Determination, is of little Importance but to convin⟨c⟩e us how ignorant we are of our own Frame, & Make.

A second Property of the Nervous Power is, That it may be greatly improved and increased by Use and Exercise.

The Powers we observe in inanimate bodies, are constant without Intension or Remission while the body is unchanged. Whether what is said of Magnetical bodies, That their Power increases, when they are kept in a certain position with regard to the Earth & other magnetical bodies that are near them, and diminishes in a contrary position, be any exception to the general Rule, I do not enquire. But it is well known what wonderfull Accuracy and | Dexterity in many muscular Motions some Persons acquire by constant Practice. This may be obeserved in almost every Craft. It is more admired in those that are less common, as in Feats of Horsemanship, of Ropedancing and Ballancing. What an accurate Adjustment of the Force, the Time, and the Order of muscular Contractions is required in such Performances, we have before observed; and it is evident that there must be an equally accurate adjustment in the Exertions of the nervous Power, by which the Muscles are contracted, since the Cause must be adequate to the Effect. And whatever Art and Skill appears in the active Exertions of the Muscles, must be ascribed no less to the Exertions of the nervous Power by which they are moved.

A third Observation with regard to the nervous Power is, That as it is capable of great improvement by the Habit of exercising it, so it is liable to be hurt and disordered, to be debilitated & even lost, by many Affections both of Mind and Body.

Fear, Anger, Suprize and other Passions of the Mind when violent, give a trepidation to the Muscles, so that we cannot use them with the same firmness and Accuracy, as when we are cool and composed. From this I think we may justly infer some involuntary and irregular Exertion of the Nervous Power. The like trepidation is sometimes produced by violent Motions of the Body too long continued, which seem for some time to exhaust the nervous Power

Although from the defects or disorders of our Muscular Motions we may for the most part infer some defect or disorder in the Exertion of the nervous Power, yet it may happen, that when this Power is duely exerted, the Nerves may be rendered unfit to convey it or the Muscles to yield to its Impression, either by external Injury or by the application of something poisonous. But when the Nerves and Muscles are unhurt every disorder or unnatural debility in the muscular Motions must be ascribed to some defect or disorder in the Exertion of the nervous Power.

This Power like all our bodily Powers is nourished by Food and may be starved by want. Fermented and Spirituous Liquors | have a very remarkable Effect upon it. When it has been weakened by want of food or by fatigue, they give some sudden temporary relief to it. But when taken to the degree of intoxication, they disorder it strangely. A Man that is drunk can hardly walk without Stumbling. He has neither that nice Sense of his ballance which he has when sober, nor can he so readily command those Motions which are necessary to recover it when in danger. If he is a Craftsman, he finds himself incapable of making good work. He blunders even in speaking & has not the Power of proper Articulation. All these Defects are owing to an improper adjustment of the muscular Contractions & probably to a disorder of the nervous Power.

Many diseases, such expecially as are seated in the Brain affect the nervous Power greatly. In some, the whole body or a part of it is quite deprived of this Power, as in Palsies & Epilepsies. In others, its Exertions are irregular & involuntary, as in Cramps Locked Jaw St Vitis Dance, & others.

Giddiness whether produced by turning round, by the motion of a Ship or Coach, or by the Distemper called Vertigo very much disorders our muscular Motions. In a fit of Giddiness a Man will throw himself to the Ground with great Violence while he thinks he is using his utmost Exertion to stand. I have seen a Man in this Distemper, sitting in his armed chair and holding it fast with both his hands, suddenly spring out of it, & fall down upon the floor or in the lap of one that sat against him There was here a great Exertion of the Muscles of Voluntary Motion. Whether the Exertion was voluntary or not may be doubted. I am apt to think that the Disease gave him a false Perception of his Ballanc⟨e⟩ so as to make him perceive things in Motion that were at rest, and himself in danger of tumbling over with the Chair above him; & that voluntarily, or by instinct, he made the Exertion which would have been proper if this had been a true Perception. |

[24] Another Cause of the nervous Power being debilitated & disordered is Old Age. Every one knows that oldage gradually diminishes Strength and Agility. But it not onely weakens, but produces a Defect, in the Adjustment of the Muscular Force that remains. Men differ greatly in muscular Strength, yet the weak can perform what is not above their Strength as perfectly as the Strong. But in Oldage even those Operations that require little strength are not performed so accurately. An old Man who has strength sufficient

Part Two: Physiology 123

to walk and to speak, very often can walk & speak onely with staggerfeet and a Stammering Tongue.

That Musical Instrument which Nature made so perfect in some, is put out of Tune. Its fine Tones are lost, & others quavering and discordant supply their Place, so that the Daughters of Musick are brought low.

I apprehend that some Defects of the external Senses in Oldage, of the Sight at least, are more owing to the Defects of muscular Motion, & consequently of the nervous Power than is commonly imagined. I have before observed that distinct Vision requires a very accurate Adjustment of three Sets of Antagonist Muscles. By one Set, both Eyes are most accurately directed towards one point. By a Second, the Pupil is contracted or enlarged according to the Degree of Light. By a third Set, the Eye is adapted to the Distance of the Object whether near or remote. In old Age the Nervous Power seems to be deficient in all these adjustments; and from this chiefly proceed the Defects of Sight in that Period of Life.

When I look on a Book whose print is rather too small for me to read without Glasses; the letters appear at first glance very indistinct; by continuing my attention for a second or two, they become more distinct, and afterwards, by alternate intervals they appear at one time more & at another less distinct. To what else can this be | owing, but that in this case, which requires all the Aid that can be had from both Eyes, I have lost the Power, of keeping both so steaddily directed to the same Object as the case requires.

One of the first Symptoms of the Decline of Sight by Age is, that they have not so distinct Vision in a faint Light as they were wont to have, and therefore are more cautious and timid in walking abroad in the night than formerly. This Defect I think is commonly ascribed to some degree of Opacity in the Chrystalline Humor of the Eye, which begins to have some tinge of Colour about middle life, & verges more and more to an Amber colour with Age. That this may partly be the Cause of the Defect now mentioned I doubt not, but it is not the sole Cause. It is certain that old Men receive less light in to the Eye in the same Circumstances, than when they were young. The Pupil is contracted, & they have lost much of that Power of dilating it, which they had in Youth. Hence, the advantage of Convex Glasses to old Men is not soley, the increasing the Refraction of the Rays, but likewise the bringing a greater

Quantity of Light in to the Pupil, which, in some Cases, may be more wanted than the other.

The loss of Power of adapting the Eye to near Objects, so as to have distinct Vision, is one of the most common defects of Sight in old Men. That this is done in early Life by the Contraction of certain Muscles, is agreed on all hands, although there be some difference among Physiologists, what the Muscles are which are destined to this Use. Whatever they be, it must be the failure of their Power that is the Cause of this Defect.

May I be permitted to mention that it was the Experience of some of these Effects of old age on the Muscular Motions that led my Thoughts to this Speculation on Muscular Motion, which, as it is owing to the Infirmities of Age, will I hope be heard with the greater Indulgence. It is both pleasant and usefull to contemplate with Gratitude the Wisdom and Goodness of the Author of our | Being, in fitting this Machine of the Body so admirably to the various Employments and Enjoyments of Life.

The Structure is admirable, as far as we are permitted to see it in this Infancy of our Being. And that internal Structure which is behind the Vail that limits our Understanding, and which gives Motion to the whole, is in a Manner most Wonderfull, though unknown to us, made subservient to our Volitions and Efforts.

This grand Work of Nature, like the Fruits of the Earth (all admirable in their kind) has its Maturity, its Decay, & its Dissolution. Like them also, in all its Decay, it nourishes a Principle within which is to be the Seed of a future Existence. Were the Fruit conscious of this, it would drop into the Earth with pleasure and Satisfaction, in the hope of a happy Resurrection. This Hope, by the Mercy of God is given to all Good Men. It is the Consolation of old Age, and more than sufficient to make its Infirmities sit light.

PART THREE
Materialism

I

2131/3/ 18 June 1774 Read Institutes of Natural & Revealed Religion. Vol
1/25, 1r 1. containing the Elements of Natural Religion by Joseph Priestly
LLD. F.R.S. J. Johnson. 1772.

In ch 1 it is mentioned as a probable opinion that as all finite things require a Cause, infinites admit of none page 8. Perhaps the Author had forgot this, when in page 12, he thinks it most agreable to Reason, though it be altogether incomprehensible by our Reason that the Creation of God is both eternal & infinite.

He thinks that all the moral Perfections of the Deity are onely modifications of his Universal & impartial Benevolence & that all Gods creatures must be happy at Last.

We cannot be certain but that Matter is capable of having even perception & intelligence added to it; though there is sufficient reason to conclude, that the Essence of God cannot be Matter.

In this first part of the Being & Perfections of God the Author seems to be very superficial. There appears no such precision as to discover a Philosophical Acumen in abstract Reasoning. There is nothing of Sentiment or beauty of Composition attempted. The work seems to be too abstract for the vulgar and too shallow for the Learned.

The Second Part of the Duty and future expectations of Mankind, is grounded upon an Analysis of the Principles of Action in the Human Mind. In which I see nothing distinct or precise. The principles of Duty he makes to be four 1 A Regard to our own happiness 2 To the will of God 3 To the good of Mankind 4 Conscience, which he says is the result of a great variety of impressions, the conclusions of our own Minds, & the Opinions of others, forming a set of Maxims which are ready to be applied on every Emergency, where there would be no time for Reason or Reflexion. This principle is as it were the result of all the other principles taken together

The peculiarities of this Volume are chiefly taken from Dr Hartley. With regard to Fatality however Priestly does not profess it openly as Hartley does but seems artfully to avoid speaking out upon that Subject.

Vol 2 pag 59 "If an unbeliever should sacrifice his fortune or his

Life in a good cause – it would give me a very high Idea of the force of good habits and mechanical Propensities in him, but a proportionally low opinion of his Understanding.⟨"⟩

5   from the preliminary Essay page 27

"Because Common Sense is a sufficient Guard against many errors in religion it seems to have been taken for granted that common sense is a sufficient instructor also, whereas in fact without positive instruction men would naturally have been mere Savages with respect to religion; as without similar instruction they would be savages with respect to the Arts of Life & the Sciences Common Sense can onely be compared to a Judge; but what can a Judge do without Evidence and proper Materials from which to form a Judgment." |

[1v] 19 June 1774 Read Observations on Man, his Frame, his Duty, & his Expectations in two Parts by David Hartley AM. 2 Vol 8° 1749

  The Author employed on this Subject 18 years, led by a Dissertation of the Revd Mr Gay prefaced to Laws Translation of Archbishop Kings Origin of Evil in which is mantained the possibility of deducing all our intellectual Pains & pleasures from association.
  Prop. 1 & 2d The Medullary Substance of the Brain is the immediate Instrument of Sensation and Motion, & the Instrument by which Ideas are presented to the Mind.
  With regard to the Sensations we have from our external Senses the Proposition is very probable. But there is no Evidence that this is the Case in the Sensations that accompany our Passions affections & other Operations of Mind in which the external Senses are not concerned. With regard to our intellectual Operations all that we know is that a Sound State of the Brain is necessary to the performance of them, & that a disorder in the brain may impede or hurt them Or perhaps improve them in one Respect while it hurts them in another. That "whatever changes are made in the medullary Substance, corresponding changes are made in our Ideas" is affirmed in the second Proposition without any manner of Evidence. The brain seems to be an intermediate Organ between the other parts of the body and the Mind. No change in the body

Part Three: Materialism

affects the Mind but by affecting the brain first. No operation of Mind produces any change in the body but by producing first some change in the brain. But as there are many changes in the body which neither affect the Mind nor the brain as far as we can discover so there are many Operations of Mind which as far as we can discover produce no change in the brain or in any part of the body. There are many operations of Mind which as far as we know do not take their rise from any change in the brain. If the Mind is a selfmoving being this must be probable and the contrary can never be proved. There is therefore reason to think, as the Author candidly acknowledges page 34 that these Vibrations of the medullary Substance are not adequate Exponents of Sensations, Ideas & Motions

That pleasure arises from Moderate Vibrations pain from violent ones.

Sleep arises from a Compression of the brain by the Blood which in that State fills the Veins & Venal Sinus's of the brain & Spinal Marrow

That Ideas are Vestiges Types or Images of Sensations.

Sensations often repeated beget Ideas; & the Vibration that accompanied the Sensation in like manner begets a Vibratiuncle which accompanies the Idea

That Sensations by being associated get such a power or Quality as that one of them being impressed, shall excite in the Mind the Ideas of the rest.

That many Simple Ideas may by Association coalese into one complex Idea which shall not appear to have any Relation to the simple Ideas as many letters coalesce into one Word & the seven primary Colours into white

Intellectual Pleasures and Pains consequently all our Affections & Passions are onely combinations of the Sensible Pleasures and Pains. |

[2r] That the first Motions of children are all automatick or mechanical, being associated with impressions made upon the Senses. That those Automatick Motions become voluntary, by being associated with that Idea, or State of Mind, (i.e. Set of compound Vibratiuncles) which we term the Will. Prop 21 Page 103

That the simple Pleasures we enjoy are many more in number than the pains we suffer. Therefore when all possible combinations are made of the Simple pleasures & Pains Pleasure will still be prevalent and in the more complex combinations will by the overballance of pleasure pain will vanish and be altogether imperceptible.

Rational Assent to a Proposition may be defined, a readiness to affirm it to be true; proceeding from a close Association of the Ideas suggested by the Proposition, with the Idea, or internal feeling belonging to the word Truth; or of the terms of the Proposition with the Word Truth. Practical Assent is a Readiness to act in such a Manner as the frequent vivid Recurrency of the rational Assent disposes us to act.

A pretended Sceptick is no more than a Person who varies from the common Usage in his Application of a certain Set of Words, viz. Truth, Certainty, Assent, Dissent &c

Prop 87 To deduce Rules for the Ascertainment of Truth & Advancement of Knowledge from the Mathematical Methods of Considering Quantity

This is done first in the Doctrine of Chances. By which we may judge of the probability of a Proposition when the Number & weight of the several Evidences for or against it are given. And in this there ⟨are⟩ two remarkable Cases 1 When the Evidences are independent and each has some Moment of its own 2 When they are dependent one upon another. An Instance of the first is when there are several witnesses to the same fact. Of the 2d When a Fact is originally testi⟨fied⟩ by the first to the 2d, then by the 2d to the 3d & so on through a certain succession of witness, each of which has a determinate degree of credibility

2 A Second Rule for the Discovery of truth is the Newtonian Differential Method. From which having given any Number of Ordinates of an unknown Curve, other Ordinates are found & the Curve described as near as the Data will permit. This may be applied to conclusions drawn either by induction or analogy.

3 A Third Rule is the Rule of False. Which may be applied to the discovery of Truth in general in this Manner. Make a Hypothesis, deduce its consequences and compare them with facts or known Truths, if they tally the Hypothesis has a certain degree of probability, the more facts it agrees with, the greater. Every thing unknown it leads to should be tried as an Experimentum Crucis. And thus the truth may be at last obtained. When the Hypothesis does not tally with known Facts or experiments made, it must be

corrected, or another made untill one is found to tally with Facts & Experiments. This Method the Author recommends as highly Usefull |

[2v] 4 The fourth Method is bring the Unknown Quantity to an Equation

5 The Method of approximating to the Roots of Equations

6 The Art of Decyphering

"These Speculations (says the Author⟨)⟩ may seem uncouth to those who are not conversant in Mathematical Enquiries; but to me they appear to cast Light and Evidence upon the Methods of Pursuing Knowledge in other Matters to sharpen the Natural Sagacity and to furnish Loci for Invention. It appears also not impossible, that future Generations should put all Kinds of Evidences and Inquiries into Mathematical Forms; and as it were reduce Aristotles ten Categories and Bishop Wilkins forty summa Genera, to the head of Quantity alone, so as to make Mathematicks and Logick, Natural History and civil History, Natural Philosophy and Philosophy of all other kinds coincide *omni ex parte*"

Pag 358 "It is the purport of this and the foregoing Section to give imperfect Rudiments of an Art of Logick which should make Use of Words in the way of Mathematical Symbols, and proceed by mathematical Methods of Investigation and Computation in Inquiries of all Sorts. Not that the Data in the Sciences are as yet in general ripe for such Methods; but they seem to tend to this more and more perpetually, in particular branches, so that it cannot be amiss to prepare ourselves, in some measure, previously"

"Logick, and Metaphysicks which are nearly allied to Logick, seem more involved in Obscurity and Perplexity than any other part of Science. This has probably been the chief Source of Scepticism, since it appears necessary that that part of Knowledge which is the Basis of all others, which is to shew wherein Certainty, Probability, Possibility, Improbability, Impossibility, consists, should itself be free from all doubt and uncertainty."

"It seems also that as Logick is required for the Basis of the other Sciences, so a Logick of a Second Order is required for a Basis to that of the first, of a third for that of the second and so on *sine limite*. which, if it were true, would, from the nature of dependent Evidences, prove that Logick is either absolutely certain, or absolutely void of all probability. For if the Evidence for it be ever so little inferior to Unity, it will by the continual Multiplication required in dependent evidences infinitely continued, bring itself down to nothing. Therefore *e converso*, since no one can say, that the Rules of Logick are void of all Probability, the *summum Genus* of them must be certain. This *Summum Genus* is the necessary Coalescence of the Subject with the Predicate. But the argument here alleged is merely one *ad Hominem*. And not the natural way of treating the Subject. The Necessary Coalescence just spoken of carries its own Evidence along with it. | It is necessary from the Nature of the Brain, and that in the most confirmed Sceptick, as well as in any other Person. And we need onely enquire into the History of the Brain, and the Physiological Influences of Words and Symbols upon it, by Association, in order to see this"

The Author distinguishes Words into four Classes 1 Such as have Ideas belonging to them, but admit of no Definition 2 Such as have both Ideas and Definitions 3 Such as have Definitions but no Ideas, of this kind are Algebraic Quantities, Scientific Terms of Art, and most abstract general Terms. 4 Such as have neither definitions nor Ideas, as the Particles the, of, to, for, but &c. He thinks Bishop Berkeley has justly observed against Mr Locke that there can be no such thing as abstract Ideas in the proper Sense of the Word Idea.

Our Passions or Affections can be no more than aggregates of simple Ideas united by Association.

II

Miscelaneous Reflections on Priestley's Account of
Hartleys Theory of the Human Mind

1 This work may be considered as an excellent Specimen of the Perfection to which the Art of Book-making may be carried, and with how small Expence of Labour & Thought a well proportioned Octavo may be produced.

2 It is still more valuable on account of the Discoveries made in this first Introductory Essay of great Vulgar Errors. Materialism has been from early Ages considedered as one of the Chief Bulwarks of Atheism Therefore while Epicurus and Hobbes and their Disciples have endeavoured to defend it, Theists and Christians have pointed their batteries against it, with great success as has been

believed by the Friends of Religion. In particular it has been thought that Dr Sam Clark has demonstrated the impossibility of Matter's being the Subject of Thought. But we learn from Dr Priestly that all was labour in vain.[a] That perception and all the mental Powers of Man are the Result of such an Organical Structure as that of the brain

Buffon, who thought as high⟨l⟩y of the Powers of organized particles as most Men, upon dissecting the Brain of an Uran Utang found so perfect a Similarity between it and the Brain of a Man, as forced him to confess that there must be something else in Man than Matter and organization. But Dr Priestly shews a firmness of Mind not to be overcome either by Buffon's Dissections or Dr Clarks reasoning

One thing however he hesitates about, whether from an Organical Structure such as that of the Brain the mental Powers of Man result *necessarily or not.*

We humbly think he could not have stopt a moment at this point, were it not that great Genius's are sometimes apt to forget their own Discoveries. For although Aristotle ta⟨u⟩ght the World long ago, that necessary Truths are onely known by Demonstration or by shewing the contrary to be impossible, & the World was so silly as to believe him, yet Dr Priestly discovered a few months ago,[b] that the proper Proof of necessary Truths is by Induction: And the evidence that any two things or Properties are *necessar⟨i⟩ly* United is the constant observation of their Union. This was a great Discovery. For it follows from it that | before Mankind had ever observed Silver to be fusible by Heat it was necessarily hard. But as soon as this observation was made; A truth which before was necessary immediately changed its nature and became contingent. Had the Philosopher recollected that Discovery of his own, he would have found no occasion to qualify this subsequent Discovery That the mental Powers of Man result from such an organical Structure as that of the brain, with that ugly alternative (whether necessary or not) How would Epicurus? How would Hobbes? How would Collins have triumphed had they lived to see this point given up to them, even by a Christian Divine?

Nor ought we to imagine that this Discovery was borrowed from Dr Hartley, to whom the Author modestly acknowledges himself

---

a. Introductory Essays page 20.
b. Introductory Observations to the Examination of Dr Reid &c pag 39

indebted for *almost* all his knowledge upon this Subject. Though Dr Hartleys Observations on Man are in his Opinion *to be ranked among the greatest Efforts of human Genius, & without exception the most valuable production of the Mind of Man.*[c] Yet in this point Dr Priestly has seen farther than even that wonderfull Man. For according to Dr Hartley[d] There is in the Human composition not onely a Gross Body and a Mind distinct from it but an elementary Body intermediate between the two. But Dr Priestly perceiving that the elementary Body & the Mind are onely an incumbrance upon the System has thrown them out.

2 Another Discovery very consonant to the first[e] is that the whole Man becomes extinct at Death. For this consession Atheists will likewise thank him as it has been one of the chief Articles of their Creed from the beginning of the World. And considering the Arguments that have been urged against it and the difficulty which even those who are most willing find in satisfying themselves upon this important point, it must give them great consolation to find their faith supported by the Authority of a christian Divine maintaining that the whole Man becomes extin⟨c⟩t at Death and that we have no hope of surviving Death but what is derived from the Scheme of Revelation If Revelation taught that the whole Man becomes extinct at Death and yet survives the grave, this we apprehend would furnish a stronger Argument against Revelation than any that Infidels have hitherto discovered. And it will need such a Champion as Dr Priestly to defend it. For there are few that can arrive at such a Strength of Faith as to believe a Contradiction. And if it is not a contradiction to be wholly extinguished at Death and yet to survive Death | it will be very difficult to say what is.

Let us suppose with Dr Priestly that all the Mental Powers of Julius Cæsar resulted from the Organical Structure of his Brain. This organical Structure is dissolved and the whole Man Julius Cæsar becomes Extinct. The Matter of this Brain however remains, but it is not Julius Cæsar for he (ex hypothesi) is wholly extinct. This Matter being capable of every form and every Organization becomes the Organized Body of a Plant, *having some degree of Sensation.* After this organical Structure is destroyed and the whole plant extinct, by a new organical Structure it becomes the

---

c. Preface to 3 Vol of Institutes p 21   d. Introductory Essays p 19
e. ibid page 20

brain of a Monkey & at last it forms the Brain of Pope Leo the 10th When all these organical Structures are destroyed let us suppose a new one produced at the Resurrection, & that Mental Powers result from it. The Question is whether this new Organical Structure be Julius Cæsar or the Plant or the Monkey or Leo the 10th or whether all these are one and the same person, and the last Production answerable for the Actions of all the preceeding. If Dr Priestly will resolve this Case it may give some new Light to his System.

This Doctrine that a Man may survive his total Extinction, and may be drawn forth out of the Limbus of Nonexistence to a second Scene of Existence is so similar to another part of the Doctors System that the one serves to illustrate the other. In his Examination of Dr Reid &c page 65. "I am surprized says he that it should have been so readily admitted that even Ideas have no Existence but when we are conscious of them. We have the same Reason to believe the Identity of an Idea as that of a Tree that of any external Body or that of our own Minds themselves. &c.⟨"⟩

In this paragraph there are two important Discoveries although the Dr confounds them together.

1 Whereas other Philosophers have held that men have no Ideas when they do not think, the Dr Finds that we have Ideas when we think nothing about them. Now if Ideas may exist without thought, it seems an easy consequence that Sensations of pleasure or pain may exist when they are not felt, & Actions when they are not acted.

Secondly we learn from this Paragraph That Ideas which were before thought to be as transient as Time itself, have a permanent Existance like Substances; that the Idea I have this moment may be the individual & Identical Idea I had yesterday, & the pain I feel to day may be the Identical pain I felt half a year ago. Whether Vibrations & Vibratiuncles, have the same permanent Nature, Dr Priestly does not inform us. Vibrations have hitherto been believed to be Successive & in a perpetual flux. And if that be so it might be expected that Ideas which are the Effects of Vibrations should not be permanent. But however this may be, Ideas retain their Identity. How long we are not told. And perhaps our Ideas Sensations and Actions are at Death embalmed and preserved during the State of Nonexistence between Death and the Resurrection, &

are then united to the new organical Structure then formed. This seems to be the onely way in which this new formed being | can have an interest in what was thought & done & suffered & enjoyed by its predecessor

Essay 2d
A General View of the Doctrine of Association of Ideas.

The Object of this Treatise being to deduce all the Phenomena of thinking from the single Principle of Association The Author in this Essay gives a brief History of the Doctrine of Association, and then applys it to resolve the Phenomena of the faculties of Memory Judgment, the Passions, the Will, & the power of Muscular Motion to which faculties he thinks all the phænomena of the Mind may be reduced.

The mechanical Association of Ideas that have been frequently presented to the Mind at the same time was, I believe, (says Dr Priestly) first noticed by Mr Locke. He next mentions Mr Gay as having attempted to deduce all our passions & affections from association. These two Authors onely are mentioned by Dr Priestly as having gone before Dr Hartley in attempting to account for the Operations of the Human Mind from association. Perhaps he did not know that Aristotle accounts for Reminiscence from this principle in the second chapter of his Book on Memory and Reminiscence. Nor that Hobbes in his Book of Human Nature has applied this principle to account for most of the powers of the human Mind; nor that Mr Hume in his Treatise of human Nature, printed ten years before Dr Hartley's Observations on Man, grounds Almost his whole System of the human Mind upon the Laws of Association. There is indeed a remarkable agreement in the Systems of Hobbs of Hume & of Hartley with regard to the Faculties of the human Mind; however widely the last may differ from the others in his Religious Principles. And I humbly apprehend that of the three Mr Hume has explained and applied the Doctrine of Association with the greatest Depth as well as with the greatest perspicuity, but of this every man will judge for himself.

It is probable that Dr Hartley ⟨w⟩as as little acquainted with Mr Humes Book, as his Compendizer, seems to be. His candor would | certainly have led him to mention it if he had. However he is very far from being of the Opinion of Dr Priestly that Mr Locke was the first who noticed the Association of Ideas.

The Doctor surely when he affirmed this had forgot what he had copied from Dr Hartley page 14, 15 "The influence of Association over our Ideas, Opinions, and Affections, is so great and obvious, as scarce to have escaped the Notice of any Writer who has treated of these, though the Word *Association* in the particular Sense here affixed to it was first brought into use by Mr Locke. But all that has been delivered by the Ancients and Moderns concerning the power of Habit, Custom, Example Education, Authority, party Prejudice, the manner of learning the manual & liberal Arts &c, goes upon this doctrine as its foundation, and may be considered as the detail of it, in various circumstances"

Had Doctor Priestly attended to this passage of his Author he probably would not have given so very superficial & partial an Account of the history of this Doctrine as he has done in this Essay, far less would he have believed that it was *first noticed by Mr Locke*.

Leaving the History of Association of Ideas, we have next in this Essay a View of the Application of it for explaining those faculties to which all the Rest may be reduced: & first that of Memory

Here Dr Priestly tells us that nothing more is necessary to explain the Phenomena of Memory, but a stock of Ideas variously associated together, so that when one of them is present, it will introduce such others as it has the nearest connexion with, & relation to.

This Account of the phenomena of Memory is as ancient as Aristotle, as was before observed. And it cannot be doubted that by things being associated together, one thing that is presented either to the memory or fancy or to the senses draws its associates along with it. Upon this principle was founded the Art of Artificial Memory which had its Teachers in very ancient Times.

Granting therefore what no body denies, That the Association of | Things in the Mind is a great Adminicle to Memory; it remains to be considered whether, this association of itself produces that faculty? To judge of this, Suppose a Cluster of Ideas brought into the Mind by association. Some of these may be Ideas of things that were, some, of things that are; some of things that will be; & some, of things which neither are nor were nor will be.

Now how shall we distinguish these? There must be a stamp put upon the Ideas of things that were, by which we shall be taught that they were, and in what order of Succession, & at what intervals of time they were. Untill this is done this bundle of Ideas produces no Memory at all.

Dr Priestly takes no Notice of this Objection to his Theory in this Essay But Dr Hartley in his Chapter on Memory[f] appears to have been aware of it. And he conceives that the Recollection of a past Fact, differs from a Reverie of Imagination partly in the Vividness of the Ideas, and chiefly in the readiness and Strength of their association.

Mr Hume had before made the Difference of Sensation Memory & Imagination to consist in different degrees of Vivacity in the Ideas ascribing to Sensation the greatest degree of Vivacity, to Ideas of Memory a less degree of Vivacity & to Ideas of Imagination the least of all. So that in this as in many other points the Systems of Mr Hume & Dr Hartley coincide.

In both these Systems, the Remembrance of an Object, the perception of it by our Senses, & the bare conception of it are held to be Operations of the same kind; they differ onely in Degree. To see an Object very faintly is to rem⟨em⟩ber it and to conceive it very strongly is to remember it. A Vibratiuncle with a little additional Strength becomes a Vibration. And a weak Vibration is, I conceive, all that is meant by a Vibratiuncle. A Man who really believed this System, if he will be consistent with himself can put no trust either in his Senses or his Memory. I⟨t⟩ suits very well with Scepticism. But a man who resolves to trust to his Senses and his Memory must reject it.

Every Man knows what Memory is, and every Man knows what is meant by vividness of Ideas or Conceptions, and their power of suggesting one another. And when we know and understand what each of these things is, we can be at no loss to know whether they are one and the same

Let every Man judge for himself whether Memory is a certain degree of Vividness in Ideas, and a certain degree of Strength | in their power of suggesting one another. To me they appear to be things quite of a different Nature & I could as easily believe that a Hat is a pair of Shoes as that Memory is a certain degree of vividness in Ideas & of strength in their association

f. page 211

A Malefactor that is going to be hanged has a *Cluster of very vivid Ideas, & very strongly associated* of what he is about to suffer, but it is not an Object of Remembrance but of foresight. If he should happen to be recovered to life after his Execution, as some have been, he will have the Ideas of the same Event probably no less vivid nor less strongly associated. But now the Event is not forseen, it is remembered. It is not the strength of his Ideas or their associations that gives him the conviction that the Event is future, in the former case. It is the force of the Laws the Strength of the Prison & the fidelity of his Guards. Nor is it in the later case the Strength of his Ideas & their associations that give him the Conviction that the Event has happened, it is his distinct Remembrance of it.

It appears evident therefore that something more than association of Ideas is required to produce Memory, & consequently that Association is not of itself sufficient to explain or account for Memory.

Judgment is next to be reduced to the association of Ideas. In order to that end Dr Priestly defines Judgment to be nothing more than the perception of the Universal Concurrence, or the perfect coincidence of two Ideas, or transferring the Idea of Truth by association from one proposition to another that resembles it.

The first part of this Definition seems to be taken from Mr Locke, & the last from Dr Hartley. But we have onely one half of Lockes definition, without any Reason shewn why the other is left out. According to Mr Locke Knowledge or Judgment is a perception of the Connection & Agreement, or Disagreement and repugnancy of our Ideas. And this Agreement or Disagreement is of four Sorts. First Identity or Diversity Second Relation Third Coexistence or Necessary Connexion Fourth Real Existence Now Dr Priestlys Universal Concurrence may tally with Lockes Coexistence or Necessary Connection; his perfect Coincidence with Lockes Identity. but the other two Sorts of Agreement to wit Relation & real Existence are left out. Does Dr Priestly think that we have no Power of perceiving the Existence of Ideas or any other Relations of them besides their Coexistence & Identity?

However this may be Dr Priestly in embracing one half of Mr Lockes Definition of Judgment seems to have left Dr Hartleys System. For if there is a power in the Mind of comparing Ideas and of perceiving certain Relations between them such as those of

Universal Concurrence, & perfect Coincidence, this Power is not that of association. For it is evident that Ideas may be associated with any degree of Strength, without being compared, without perception of any Relation between them. The first of these powers may be supposed in any degree, without the others. And therefore the Power of Association does not account for the powers of comparing our Ideas and perceiving their Relations.

Dr Hartleys Definition of Assent & Dissent, that is of Judgment, seems to tally better with his System. He makes them to be,[g] Those very complex internal feelings which adhere by association to such clusters of words as are called Propositions. Whether this Definition, or that which Dr Priestly has substituted in its place, be the most perspicuous & the most accurate, every one may judge for himself.

In accounting for the Passions from Association Dr Priestly seems to allow that we have originally a desire of Pleasure and aversion to pain which do not arise from association. These therefore are original Principles In accounting for our other principles of Action from these original ones & the Doctrine of Association, the same things are said which were said by Epicurus of Old & have been said by all those who have defended the selfish System or ancient Epicurean System with regard to human Actions; without taking any Notice of what has been said on the other side by the best ancient Philosophers, or by Shaftsbury Hutcheson Butler and many others among the Moderns.

Dr Priestly acknowledges that according to his System all our Passions are at first Interested, and that the Disinterested are got by Association in the same Manner as some Men get the Love of Money for its own sake.

We acknowledge that the Love of Money for itself is got by the habit of associating it with other things which we desire & which it may be a mean of procuring. All the world acknowledges this. And what is the Consequence? It is that every Man despises this passion in others | and endeavours to justify himself from the imputation of it. The greatest Miser perswades himself that he either does, or will reap benefit by increasing his wealth, And professes to put no value upon it on its own Account.

If other disinterested passions were got by the same means &

g. page 158

owing to like associations we should have the same Reason to be ashamed of them in our selves & to despise or pity them in others. The Love of Parents, Children, Friends, Country, nay of the Supreme Being himself would upon this System be the weaknesses of human Nature, and the effect of Associacions which pervert the Judgment & make us mistake the means for the End. The onely Perfect Man according to this System would be the perfectly selfish Man, who has no desire but for his own Pleasure, or other things as means to that End; no aversion but to his own Pain, or to other things in as far as they may procure his pain.

The next Operation of the Mind which Dr Priestly is to explain from association is Volition, which he tells us is a modification of the passion of desire. Mr Lock, on whose Principles Dr Hartley founded his System as Dr Priestly tells us, has taken pains to refute this vulgar Error, & has shewn very clearly that Volition is not desire nor a Modification of Desire. Yet as if Lockes Judgment & Reasoning were unworthy of the least consideration Dr Priestly without offering any proof, takes it for granted that Volition is a Modification of Desire.

What follows of this Essay, is what we apprehend the Author would have us take for an Explication of the Power of Muscular Motion from the Doctrine of Association. The purport of it is to shew, or rather to affirm That those muscular Motions which have most the Appearance of Instinct are at first Automatick & involuntary and afterwards become voluntary by means of Association.

Dr Priestly seems to have a great Disgust to Instincts. There is something in them that offends his Taste, he thinks them an operose and inelegant Contrivance, and therefore does not | allow that there is any such thing in the frame of Man or of other Animals. What has most the Appearance of Instinct, he thinks, has been in a Manner demonstrated by Dr Hartley to have been originally Automatick

To what Association of Ideas this Aversion to Instincts has been owing in Dr Priestly he can best explain. It would not however have been improper, when he has said so much against Instincts, to have told us what Instinct means. But his Zeal against instincts will not allow him to do any thing but abuse them.

By Instinct in Animals is commonly meant a Propensity to certain Actions, which is neither the Effect of any rational Motive nor the Effect of Habit, but of the Constitution of the Animal.

Such Instincts have been believed to be both in Man & in brute Animals, especially in the beginning of Life when the Animal has no knowledge nor Experience of what may be for its good, or for its hurt; and therefore to supply the Defects of Reason and Experience, has need of some inward Monitor and Prompter, to lead it to those Actions which are necessary for its own preservation or the continuance of the kind.

It has been thought both in ancient and Modern times that the Instincts of Animals furnish one of the clearest & most cogent proofs of Contrivance and Design in their Frame, and consequently of the Existence of a Wise Author of Nature.

Dr Priestly's Zeal against the Existence of Instincts seems the more surprizing that his Author Dr Hartley acknowledges their Existence and thinks there is no difficulty in reconciling them to his System. He supposes the Bodily Frame in Brutes to be so formed that miniature Vibrations spring up ⟨in⟩ it at certain ages and Seasons of the Year, that the Ideas & voluntary Motions corresponding to these Vibrations must of consequence attend them, & mix themselves with impressions and acquired Ideas; that this is a kind of inspiration to Brutes, helping out that part of their faculties which corresponds to Reason in us[h] |

Philosophers who agree in the Existence of Instincts may yet differ with regard to their immediate Cause. Some with Dr Hartley may ascribe them to the Original Frame of the Animal Body others with Malebranche may ascribe them to impulses given immediately by the Deity as there is occasion for them. And others may modestly acknowled⟨g⟩e their ignorance of the Cause although they perceive such manifest marks of Contrivance and Design in the Effect as lead them to believe that it must, either mediately or immediately, proceed from a wise & intelligent Cause.

The title of the third Preliminary Essay is, Of complex and abstract Ideas. But the Intention of it seems to be to perswade us That all the Ideas which Mr Locke calls Ideas of Reflexion such as those of Mind, Thought, Judgment &c, are in reality compositions made up of the Ideas of Sense. This Notion seems to agree well with the System of Materialism which Dr Priestly has adopted and has been maintained by other Materialists before him particularly by Epicurus and by Hobbes. And indeed it seems reasonable to

h. page 246

Part Three: Materialism

think if a certain composition of Matter can, whether necessarily or not, produce Thought Judgment & reasoning, that a certain composition of the Ideas of Matter may produce the Ideas of thought Judgment & Reasoning. This is an Argument in Favour of Dr Priestlys System which Seems to have escaped his penetration, & for which we hope to receive his thanks. We may expect this the more because his Arguments upon this point seem to require some reinforcement.

For after proposing his proposition & acknowledging that it is not very easy to conceive how intellectual Ideas can be composed of sensible ones, he proceeds to offer some considerations to lessen this difficulty a little. Now we apprehend that if the difficulty of conceiving a Proposition should be not onely lessened a little but totally removed, it will not follow that the proposition is true. The Arguments that prove its truth must be something else than considerations that facilitate its conception. |

[12] To facilitate this conception he tells us That a whole Group of Ideas shall so perfectly coalesce into one as to appear but a Simple Idea.

The instance which he thinks comes nearest to this is That a mixture of the several primary colors produce white. But this instance though the nearest to the case before us is too distant from it to yield any solid argument. For he ought to have shewn that the Idea of White is compounded of the Ideas of the primary colours. Or that whatever may be affirmed of the Colours may be also affirmed of their Ideas. If the last is true, as it is evidently supposed in Dr Priestlys Argument, it will bring to light many Classes of Ideas that have escaped the Observation of Philosophers. For if what ever is affirmed of Objects may also be affirmed of their Ideas, we shall have blue & red & green Ideas Ideas that are weighed by the ton, & others that are measured by the bushel, elastick and unelastick Ideas, animal & vegetable Ideas & a thousand other kinds.

If it be absurd to ascribe to Ideas Color & Weight & Elasticity, & indeed if this is not absurd it will be hard to say what is, then Objects may have Qualities & Relations which their Ideas have not, and a mixture of primary Colours may make a white Colour, although a mixture of the Ideas of the primary Colours do not make the Idea of white. Dr Priestly might as well have argued that because several Metals in fusion mix and coalesce so as to appear but one simple Metal, therefore the Ideas of those Metals may coalesce into one simple Idea. And by the same kind of reasoning because Metals are malleable & fusible, it will follow that their Ideas are malleable & fusible. Such is the Reasoning which we are to receive for a Proof that a whole group of Ideas shall so perfectly coalesce as to appear but a simple Idea.

But to illustrate this doctrine farther & facilitate its conception the Author gives instances of some Ideas formed in this Manner. Such is that of a Player. A child has seens a company Act on the Theatre in a great variety of Characters, & is told that he must call them players. That word will excite in his Mind an Epitome as it were | of all that he has seen them perform. Even the Features and most striking gestures of the principal Performers will be conspicuous in it. And by degrees as all these particulars get intermixed, and compleatly associated, whatever belonged to the separate persons will be dropped, and something will remain annexed to the Term, that had been observed in them all. This says Dr Priestly is the process called Abstraction, & it is by means of this process chiefly that we acquire those Ideas which have been referred to Reflexion.

We have no Exception to this Account of the Manner in which a child may learn to fix a distinct meaning to the word Player. It is no doubt by observing in course of time what is proper to individual players and distinguishing that from what is common to them all, but two things here deserve to be noticed.

First that the Idea of a Player is not formed by the Child, by means of association, but by a contrary Operation of distinguishing and separating the things common to all players from those that are proper to each. And when this Idea is formed and made distinct, it is not a mixture and combination of the Ideas of all the Individual players, it is a selection of what is common to them all, and therefore in reality more simple than any one of them. The Idea of a Player is more simple & less complex not in appearance onely, but in reality than the Idea of Roscius or of Garrick. And the Examples brought by Dr Priestly in order to shew that very complex Ideas, though to appearance simple, may be formed by combination & association; show onely on the other hand, that Ideas very simple in reality as well as in appearance may be formed from those that are complex by division and Separation.

Part Three: Materialism 145

Secondly if it is by Abstraction chiefly that we acquire those Ideas which have been referred to Reflexion; It will follow that the Ideas of Reflexion are more Simple than those they are abstracted from instead of being more complex as Dr Priestly affirms.

[14] *In the same Manner*, says Dr Priestly, *that we get the Idea | annexed to the word Player, we get the Idea that we have to the Word Thought or Thinking; which, in fact, is an abridgement or coalescence of the various external Signs or marks, and also of the internal feelings by which exclusive of the outward form, a Man is distinguished from a brute Animal*

In this Account of thought there is indeed a very strange assemblage of Ideas but very little Coalescence. We may observe from it First that Thought or thinking distinguishes a Man from a Brute Animal. Yet this same Author maintains that Brute Animals Think & Reason in the same Manner that Men do. Secondly we learn that the Idea of thought is composed of external marks and Signs & internal Feelings. Thirdly that it is an abridgement or coalescence of these. Indeed it is as likely that Abridgment should be Coalescence as that Association should be Abstraction. Fourthly we may infer from this curious passage that one who never saw a brute Animal can have no Idea of Thought or Thinking. These are curious Discoveries and give much Light to the Idea of Thought.

We would be glad to be informed by Dr Priestly whether a Man, when he thinks, is not conscious of his thoughts? Whether he has not the power of reflecting upon his own thoughts, & making them an Object of Thought? Whether such Reflexion will not give him the Idea of Thought, although he had never considered the various external Signs or Marks, & also the internal feelings by which a man is distinguished from a Brute.

But I am weary of Pursuing the incoherent & absurd Notions of this flimsy writer, who surely mistook his Talent when he attempted to write upon abstract Subjects |

[15] Dr Hartleys *Observations on Man, his Frame, his Duty, & his Expectations* in two Vol 8° were published Anno 1749 so that the Publick has had sufficient Time to form a Judgment of its Merit. The Candor & Piety which shine through the whole work; as well as the very extensive acquaintance with the several Branches of the Medical & Mathematical Sciences, which he discovers, give him a just Title to our Respect, however we may differ from him

in Opinion. Conjectures upon dark and dificult subjects modestly proposed, are no proper Subject of Criticism. Like Queries they neither affirm nor deny, but suggest Matter to future Enquirers. Dr Hartleys Book consists mostly of such Conjectures & consequences drawn from them. It is therefore no Wonder, that, as by Dr Priestlys Account it seems to have been little read, so it should be less criticised.

The Opinion which Dr Priestly has formed of it seems to differ no less from the Authors own Opinion than from that of the Publick. He is "astonished that any person should write upon the Subject of the human Mind, without taking Notice of so capital a Performance as that of Dr Hartley; who beginning where Locke left it, has raised a System that is equally amazing for its Simplicity & extent. For my part, says he, I do not hesitate to rank Hartleys Observations on Man among the greatest Efforts of human Genius; and considering the great importance of the Object of it I am clearly of opinion that it is, without Exception the most valuable production of the Mind of Man. – He has no doubt but the time will certainly | come, when the general Prininciples of Hartley as well as of Locke, will be fully established, and when every contrary Hypothesis will be forgotten. If they be remembered at all, it will be with astonishment, that, appearing after such a Work as Hartleys, the least attention should have been given to them". (Pref. to 3d Vol of Institutes &c) He thinks that "Hartleys System of Vibrations has been Demonstrated by him in as Satisfactory a Manner as can be expected, in a Subject so very obscure as this necessarily is". Essay 1s⟨t⟩ page 10. That he has "thrown more usefull Light upon the Theory of the Mind, than Newton did upon the Theory of the Natural World⟨"⟩ (Examination of Dr Reid &c pag 2d)

If the general Principles of Hartleys System be fully established, it is impossible that those of Locke can. They contradict each other in many capital Points, and he must be extreamly ignorant of one or the other who does not perceive this. But to pass this, it is pleasant to observe the contrast between Dr Priestly's opinion of this work and the Authors, expressed in his preface. "If the Reader, says he, will be so favourable to me as to expect nothing more than Hints & Conjectures in difficult & obscure Matters, and a short detail of the principal Reasons and Evidences in those that are clear, I hope he will not be much disappointed". And in his

Introduction after mentioning the proper Method of Philosophizing, by Synthesis & Analysis, recommended & followed by Sir Isaac Newton he adds "I shall not be able to execute, with any accuracy, what the Reader might expect of this kind, in Respect of the Doctrines of Vibrations and Association, and their general Laws, on Account of the great Intricacy Extensiveness and Novelty of the Subject. However I will attempt a Sketch, in the best manner I can for the Service of future Enquirers."

If we could suppose the Author to have had the same Opinion of his Theory which Dr Priestly has; his apparent Modesty must be a silly Affectation inconsistent with Sincerity. But it would be very injurious to Dr Hartley to think so. We believe | that he was candid as well as modest; and that what he really meant to propose as Hints & Conjectures has been mistaken by Dr Priestly for Demonstration. Whether the Panegyrick Dr Priestly makes upon this Work shall contribute to raise its Esteem or to lessen it, Time will discover. It must be granted that Books as well as Men may suffer by injudicious Praise no less than by Detraction

Dr Priestly thinks "that Hartleys Observations on Man could not have failed to have been more generally read and his Theory of the human Mind to have prevailed, if it had been made more intelligible, and if the work had not been clogged with a whole System of Moral & Religious Knowledge; which, however excellent, is, in a great measure, foreign to it – That many excellent Articles, he thinks he may say all the Articles, have been, in a great measure, lost to the World, in consequence of being published as parts of so very extensive a System." (Preface) To remove those two obstacles to the Success of Hartley('s) System is the professed design of Dr Priestly in this mutilated Repubblication which he calls Hartleys Theory of the Human Mind.

If Dr Priestly thought it necessary to give Reasons to the Publick for his making a new Book, we think he has been unlucky in the present case; and that he detracts from his Author very unjustly in order to give importance to himself.

We see no Reason to charge Dr Hartley's Observations on Man with being in any Degree unintelligible. His Words are well chosen. His sentences easy perspicuous and unaffected, and the Whole Book pertinent to the Subject, & digested in good Order and Method. When he uses new words he explains their meaning; and

[18] in every part of his work he appears to have a distinct meaning and to | express it in a proper Didactick Style. He frequently borrows Arguments or Illustrations from Medical or Anatomical Facts, or from Mathematical Truths. If these are not understood by some Readers, this ought not to be imputed to the want of intelligibility in the Book, but to Ignorance in the Reader.

But the Method which Dr Priestly has taken to remove this Obstacle and to make his Author more intelligible is new and Curious, and may, if the Pattern is followed, facilitate the work of future Commentators. He ommits what he thinks proper to ommit, and makes a Book of what remains, very rarely altering or adding a Single word. And this is his Method of making the Author more intelligible.

The Custom heretofore has been to make an Author more intelligible by adding Notes or Commentaries to the Text. Dr Priestly illustrates his Author by taking away a great part of the Text itself, & leaving the Remainder as it stood before. He taught us in his third Essay that very complex Ideas (such as that of thought) are formed by an Operation called Abstraction. And now we learn that an Author less intelligible, is made more intelligible, by an Operation not very unlike the former, commonly called Subtraction These are new discoveries & very consonant to each other. We confess however we are yet to learn how this mutilated Edition of Hartleys Work contributes to make it more intelligible.

But Dr Priestly had another Obstacle to remove in order to make Hartleys theory *to be more read*, & *to prevail*, namely that *it is clogged with a whole System of Moral and Religious Knowledge which however excellent is in a great measure foreign to it*. A most lamentable thing indeed it is that an | excellent System of moral and Religious Knowlege deduced from a just Theory of the human Frame should prove a Clogg to that Theory, should hinder it from being read and prevailing, and like a dead Weight be ready to sink it in the sea of Oblivion We shold be glad to be informed by Dr Priestly when this Æra commenced, wherein the World began to have so great contempt for Moral and Religious Knowledge and to put so high a Value upon Metaphysicks. Dr Hartley surely valued his Theory chiefly on this Account, that he thought it subservient to moral and Religious Knowledge. And no man who has the least Tincture either of Morals or Religion will think its application to serve those Ends to be a Clogg upon it. Dr Priestly thinks moreover *that this Theory was in a great measure lost to the World by being published as a part of so very extensive a System.* A most unreasonable and most unaccountable World surely it must be, if a System is lost to it because it is extensive. If real Knowledge be of any Value, the farther it is extended, the greater must its value be. We cannot be perswaded to entertain so very contemptuous an Opinion of the World as to think, that a System built on a Solid foundation and properly reared should be lost to the World by this circumstance of its being extensive, which ought to enhance its Merit: or that it should be less read and less regarded on account of its containing an excellent System of Moral & Religious Knowledge.

Dr Hartleys work, whatever we may think of the Solidity of its foundation, must be acknowledged by all who are acquainted with it, to have great Regularity in its Structure and a close Connection and Dependance in its Parts |

The work corresponds perfectly with the Title; there is an Unity of Design in the Whole, and the Materials are disposed in the most proper Method. We cannot help thinking therefore that by being dismembered in this Publication, as it has gained nothing in point of solidity or Perspicuity, so it has lost much in point of Beauty and Symmetry. |

We beg leave farther to observe that Hartleys Theory of the Human Mind and its operations, is, of all the Parts of his System, that which is most difficult to defend. While he speaks of the Vibrations and Vibratiuncles of infinitesimal Particles, and of the Operations of an Elastick Ether which pervades all bodies without sensible Resistance, it is equally difficult to prove his conclusions and to disprove them. It is like giving us a Theory of the Characters and Manners of the Inhabitants of Jupiter, who are not more removed from human Knowledge by their distance than the infinitesimal particles of bodies by their minuteness But when he forms Conclusions concerning the Operations of the human Mind, the Objects are within our reach, and those who have by practice acquired the habit of reflecting upon what they are dayly conscious of, have a touchstone whereby they may discern what is genuine from what is counterfeit

One cannot without Reluctance criticise a work written with so much Candor & so good Intention as that of Doctor Hartley, & proposed as Hints and Conjectures for the benefit of future Enquirers. Whether his Disciple Dr Priestly by proposing it to the World as a System almost demonstrated may provoke any Examination of it we cannot say. But if any one should undertake this task, the Censure of whatever may be found weak or ill founded in it, ought in Justice to fall upon the Dogmatical Spirit of the Disciple & not upon the Master. We desire that this Caution may be applied to what has been before said upon Dr Priestlys second & third Essays, wherein he gives an Idea of Dr Hartleys Theory of the Mind, in order, as he says, to show the possibility of it, and to be usefull to those who are to enter upon the Study of it. And we proceed to make a few observations upon that part of Dr Hartleys work which is before | us with the Respect which we think truly due to an amiable worthy and Ingenious Man.

1 We see in Doctor Hartley a great Degree of Invention in framing Conjectures to account for the Phænomena of Nature, & a great Degree of Ingenuity both in answering objections against them and in drawing from very distant corners out of his extensive fund of Knowledge a variety of instances to illustrate & confirm them. And although he proposes them with a Modesty truly amiable yet they seem to have taken a stronger hold of his own Judgment than Conjectures ought to do. To give an Instance of this; he tells us in his preface that he was not aware that the Doctrine of Necessity followed from his System till several years after he had begun his Inquiries; nor did he admit it at last without the greatest Reluctance. Yet the love of the System seems to have at last so far overcome this Reluctance as to lead him to embrace this doctrine and even to embrace those Consequences of it which indeed recommend it to the enemies of Religion, but which the friends of Religion who have been necessitarians, have generally rejected with horror, and have laboured with all their might to shew not to be justly drawn from their Doctrine. See how he answers a common Objection against the Doctrine of Necessity Hartleys Theory page 343. "Eightly, it may be said, that the Doctrine of Mechanism makes God the Author of Sin. I answer, that till we arrive at self annihilation, sin always will and ought to appear to arise from ourselves; and that when we are arrived thither, sin and evil of every kind vanish." It seems then that before we arrive at self annihilation we

Part Three: Materialism 151

ought to believe that sin arises from ourselves; why then does Dr Hartley exert the whole Strength of his Genius to perswade us to believe the contrary of what we ought to believe. Surely he did not think that his book would never be read by any but those who have arived at self annihilation.

[23] Nor can we think that he believed that the embracing his Doctrine of the Mechanism of the Mind would of it self bring | a man up to that State. The strongest Advocates for Necessity have acknowledged that it may be abused by bad Men to make them worse and more abandoned. Even Mr Hobbes from an Apprehension of the bad Use that might be made of this Doctrine, desires that his treatise in Defence of Necessity, one of the best that has been wrote on that side of the Question, might be communicated onely to the Marquis of Newcastle his Patron & to the Bishop against whom it was wrote. And one might have expected that Dr Hartley would have reserved this Doctrine for the Adepts who have arrived at Self annihilation, since none else ought to believe it.

What follows is still more strange. "That when we arrive at the State of Self annihilation Sin and Evil of every kind vanish." That is, when we arrive at this State our eyes will be opened and we shall see that those Actions of ours which before we took to be Sins and very culpable Evils, were really the works of God and no Evils at all. Was there ever any thing said by a Hobbes a Colins or a Bolinbroke, which more directly overturns all the Moral Attributes of the Deity and all Distinction between Virtue and Vice! This is a lamentable instance to what length a prediliction for a favourite System may carry a good and a pious Man; for such we take Hartley to have been.

Although Hartley was well acquainted with the writings of Sir Isaac Newton & had a just admiration of them, yet he seems to have payed very little regard to that chaste Method of Philosophizing which that great Man strictly adhered to and which was before pointed out by Lord Bacon as the onely Method by which we can acquire any Knowled⟨g⟩e of Natures works. Conjecturing & Philo-
[24] sophizing are quite different | things. The first is left free from the restraint of Rules. But the Last is restrained by Strict Rules; and is allways to be esteemed spurious & illegitimate when it goes beyond them. The necessary consequence of this is, that in a philosophical Work which pretends to unfold the Causes of the Phenomena of

Nature or to establish any new Principles. The things that are proved legitimately ought to be kept distinct from those that are conjectured, or built upon slight probabilities. Sir Isaac Newton has given an Example worthy to be followed; In the third Book of his Principia which contains a System of Knowledge with regard to our Planetary World. Every part of his System, which he deliv⟨er⟩s in a Chain of Propositions with Corolaries, is deduced by strict reasoning from the Phenomena according to the Laws of Induction Sometimes he mentions his Conjectures in Scholia, which are reserved for them. But no Conjecture is ever set down as a Proposition or Corolary by him, Nor is there ever any subsequent Proposition or Corollary grounded upon these Conjectures. In like Manner in his Opticks, he goes on in a Regular train of Propositions founded on a just Induction from Experiments as far as he had Experiments to support him in this strict Method. When he comes to that part of his Subject in which he had not a sufficient Number of Experiments for a compleat Induction he sets them down under the Title of Observations, and last of all come his Conjectures in the modest form of Queries, in which nothing is affirmed or denied. So cautious was that great Philosopher of adulterating his noble Discoveries by mixing them either with lame inductions or with his Conjectures.

Dr Hartleys work consists of a Chain of Propositions and their Corollaries, digested in good Order, and in Scientifick form | He acknowledges that a great part of them are conjecture, and Hints onely; yet these are mixed with the propositions legitimately proved, without distinction; corollaries are drawn from them & other propositions grounded upon them, which all together make up a System. A System of this kind resembles a Chain which has some links aboundantly strong and others very weak. The Strength of the Chain is not to be measured by that of the strong links but by that of the weak, for when they break, the chain falls in pieces.

That this is a just Account of Dr Hartleys System, we may be satisfied not onely from his own Acknowledgement in his Preface before quoted; but from the Book it self. Nor need we go farther than the first two Propositions.

The first Proposition is "That the white medullary substance of the Brain, Spinal Marrow, & the Nerves proceeding from them is the immediate Instrument of Sensation & Motion." If we take *Sensation* to mean, as Dr Hartley defines it "those internal feelings which arise from the impressions made by external Objects upon the several Parts of our Bodies"; the proposition is undoubtedly true and is sufficiently proved, as he observes, in the writings of Physicians and Anatomists. We have no exception therefore to any thing that can be deduced from this Proposition by just Reasoning.

The second Proposition we apprehend is mere Conjecture It is "That the white Medullary substance of the brain is also the immediate Instrument by which Ideas are presented to the Mind: Or, in other Words, whatever changes are made in this Substance, corresponding changes are made in our Ideas and *vice versa*". It is here to be observed that Hartley distinguishes Ideas from Sensations in the same manner as Mr Hume distinguishes them from Impressions. Taking Ideas in this Sense, how do we know that no Idea is presented to the Mind by Memory or Imagination but in consequence | of a corresponding change of the Medullary Substance? We grant that there are some changes in the Brain which affect the mental Powers, and that a certain sound State of the Brain is necessary to the Exercise of them all. Thus far the dependance of the Mind upon the Body and particularly upon the medullary Substance of the brain is sufficiently evident and acknowledged by all.

It must be acknowledged at the same time that we know not what that State of the Brain is which we call a Sound State, and which is necessary to the exercise of our Mental Powers; nor do we know what are the changes from this sound State by which Memory Reason or any other Mental Power is injured. There may be changes in the Brain that hurt our faculties without producing any Ideas that correspond to those changes. There may be changes in the brain that are hurtfull to health without hurting the mental faculties or producing Ideas, & there may be changes in the brain which are attended with none of those effects.

On the other hand although very intense thinking or violent passions of the Mind may sometimes produce changes in the Brain: it by no means follows that every Idea that passes in the Mind has a corresponding change in the brain which occasions it. This proposition therefore must be considered as one of the Hints and Conjectures of which the Author speaks; yet it is the main pillar of his System. For from this proposition it follows that the operations of the Mind are all governed by the Laws of Motion, which produce the various changes in the medullary Substance of the Brain, and on this System Hartley very properly calls them all Mechanical. It was probably this notion, of the Mind being indebted to the Medullary Substance for every thought & every operation belonging to it, that led Dr Priestly into Materialism, & brought him to think the Mind an Burthen & incumbrance upon Hartleys Scheme. Indeed it led Hartley himself to the borders | of Materialism; but he stopt short, and left to Dr Priestly the honour of removing this Incumbrance of a Mind, and by that means *of helping much*, as he says, Hartleys Hypothesis

Philosophy has been in all Ages adulterated by Hypotheses, that is, Systems that are built partly on facts but much upon Conjecture. 'Tis pity that a Man of Dr Hartleys knowledge and Ingenuity should have followed the Multitude in this Tract, after the futility of it had been so clearly pointed out by Bacon & Newton. The last considered it as a reproach when his System was called his Hypothesis, and says with disdain of such imputation, *Hypotheses non fingo. Quicquid enim ex Phænomenis non deducitur Hypothesis vocanda est; et Hypotheses, seu Metaphysicæ seu Physicæ, seu Qualitatum Occultarum, seu Mechanicæ, in Philosophia Experimentale locam non habent.*

It is indeed very strange that Dr Hartley should not onely himself follow a Method of Philosophyzing so much condemned by Newton & Bacon, but that he should instruct others in their Enquiries to follow this Method. For so he does (Prop 39 Hartleys Theory) in that Part of the Proposition where he deduces Rules for the ascertainment of Truth from the Rule of Position in Arithmetick

III

The Modesty and Diffidence with which Dr Hartley offers his System to the World, by desiring the Reader *to expect nothing more than Hints & Conjectures in difficult & obscure Matters* and a short detail

A Second Condition required in the Causes of Natural Phenomena assigned by Philosophers is that they be sufficient for explaining the Phænomena Let us consider how far Vibrations of the Nerves & Brain answer this Condition when brought to explain the Phenomena of Sensation & perception.

It would be doing Injustice to Dr Hartley to conceive him a

Materialist. He always proposes his Sentiments with great Candor and they ought not be carried beyond what he expresses. He thinks indeed that it follows from his Theory, that Matter if it could be endued with the most simple kinds of Sensation might arrive at all that Intelligence of which the human Mind is possessed; and that his Theory overturns all the Arguments that are usually brought for the immateriality of the Soul from the Subtlety of the internal Senses and of the Rational Faculty. But he does not take upon him to determine whether Matter can be endued with Sensation or no, and therefore would not be any way interpreted so as to oppose the immateriality of the Soul. He even acknowledges that Matter & Motion however subtly divided & reasoned upon, yield nothing more than Matter and Motion still.

It would therefore be unreasonable to require that his Theory of Vibrations should *account* for Sensation & perception in the most proper Sense of that word. It would be ridiculous in any man to pretend that thought of any kind must be the necessary consequence of Motion, or that a vibration of the Nerves must cause Sensation any more than the Vibration of a Pendulum. Dr Hartley disclaims this way of thinking & therefore it would be unjust to impute it to him. All that he pretends is that in our constitution there is a certain connexion between Vibrations of the Medullary Substance & the thoughts of the Mind so that the last depend entirely upon the first, and every kind of thought in the Mind arises onely in consequence of a Corresponding Vibration or Vibratiuncle in the Nerves & Brain. Our Sensations arise from Vibrations and our Ideas from Vibratiuncles or Miniature Vibrations. And it is to be observed that under the Word Sensation he comprehends the perception of external Objects, & that all the Operations of the Understanding are by him comprehended under the two terms of Sensations & Ideas.

What may be justly required in the Theory of Vibrations, when brought to account for the Phenomena of Sensation and Perception is that the different kinds and degrees of Vibrations should correspond to the different kinds & degrees of Sensation & Perception. But they by no Means do so. For consider onely our Sensations We have five Senses whose Sensations differ totally in kind. By each sense, excepting perhaps that of hearing, we have a variety of Sensations which differ not in degre onely but in kind. How many tastes and Smells are there that are specifically different, each of them being susceptible of innumerable degrees of strength and weakness. Heat & cold. roughness & smothness hardness & softness, pain & pleasure, are things totally different in kind, having each of them an end less variety of degrees. Sounds have the qualities of acute & grave strong & weak in inumerable degrees, & the varieties of colour are many more than we have names to express. How shall we find variations of the Vibrations corresponding to all this Variety of the Sensations? We know that Vibrations have two Qualities that differ in kind they may be quick or slow, they may be strong or Weak. and Each of these qualities may be varied without end in their degree. Dr Hartley adds two other Qualities, to wit, that they primarily affect one part of the brain or another, and they may vary in their Direction according as they enter by different external Nerves. But these seem to be added to make a Number. For as far as we know Vibrations in an Elastick uniform Substance spread over the whole, & in all directions However taking in all those four Variations of Vibrations. should it not follow that we may have by them four Senses, and onely one kind of Sensation by each varying in degree but not in kind. That we have one Sense arising from the quickness or Slowness of the Vibrations another from their Strength a third from their place & a fourth from their direction? But we have five senses & by each of them we have Sensations more than sufficient to exhaust all the variations we can conceive of Vibrations, including even the Vibratiuncles. Dr Hartley indeed seems to be sensible of the difficulty of finding

[2r] Vibrations to suit all the Variety | of our Sen⟨s⟩ations, & strains very hard not onely by heaping supposition upon supposition, conjecture upon Conjecture not onely by calling in all the Phisiological & Pathological Observations his extensive knowledge furnished him with to give some credibility to his Hypothesis. In find⟨ing⟩ Vibrations to correspond to one of the Senses he seems to forget that those must be excluded which are appropriated to the other Senses

Philosophers have accounted in some degree for our various Sensations of Sound by the Vibrations of elastick Air. But it is to be observed First that we are certain such Vibrations do really exist. Secondly they tally exactly with the Phenomena of Sound. We cannot indeed shew how any Vibration should produce the Sensation of Sound, this must be resolved into the will of God or into

Part Three: Materialism

some cause altogether unknown. But we know that as the Vibration is strong or weak so the Sound is loud or low. We know that as the Vibration is quick or Slow so the Sound is acute or grave. We can point out that relation of Synchronous Sounds which produces harmony or discord and that relation of Successive Sounds which produces Melody. And all this is not conjectured onely but proved by a sufficient Induction of Experiments it describes to us the Work of God not the conjecture of Men. The Account of Sounds therefore is so far Philosophical, although perhaps there are many things relating to Sound that we cannot account for, & whose causes remain to be discovered. If such an account could be given of other Sensations By Vibrations of the medullary Substance, it would deserve its place in Philosophy. But when we are told to account for them by Vibrations of a Substance which no man could ever shew to be elastick nor to have vibrations. When such vibrations supposing them to exist cannot be shewn to tally in any degree with the variety & different Natures of the Sensations they are supposed to excite. Such a System deserves no place in Philosophy. It is not the Work of God that is described but the imagination of Man

But how can we expect any proof of a Connexion between Vibrations & Thought when the Existence of such Vibrations was never proved. The proof of that Connexion cannot be stronger than the proof of this it is acknowledged that we could not infer the existence of thought from the Existence of Vibrations; & it is as evident that we cannot infer the existence of Vibrations from the Existence of Thought. The Existence of both therefore must be known before we can know that they are connected. As to the Existence of our Thought we know it by Consciousness, a kind of Evidence which never was called in Question. But as to the Existence of Vibrations in the Medullary Substance of the Nerves & Brain no proof has yet been brought.

All therefore that we have to expect from this Hypothesis is that Vibrations considered abstractly should ⟨seem⟩ capable of like variations in kind and in degree as the thoughts they are brought to account for. So that for every particular kind of thought, a kind of Vibration may be imagined that corresponds to it, & the divisions & subdivisions of thought may run parallel with the divisions and Subdivisions of Vibrations. This is the very least that ought to be expected to make the Hypothesis plausible. But we do not find even this in the Hypothes⟨is⟩ of Vibrations For to omit all those thoughts & Operations of Mind which the Author calls Ideas & which he accounts for by Vibratiuncles; To omit the perception of External Objects which he comprehends under the Name of Sensation we shall confine our Selves to Sensations properly so called, & even of these we shall confine our selves to the Sensations we have by our External Senses, omitting all the Sensations which accompany our passions and Affections.

There is a figurative Meaning of the Phrase of Impressions upon the Mind, which we took notice of in the Observations made on that Word. But this meaning applies onely to objects that are interesting When I see an Object with perfect Indifference I apprehend it is not good English to say that it makes an impression upon my Mind.

There is another Conclusion which has been drawn from the Impressions made on the Brain in perception, which I conceive to have no Solid foundation though it has been adopted very generally by Philosophers. It is that the impressions made on the brain are images of the Object perceived & that the Mind being seated in the Brain as in its Chamber of Presence, as Mr Locke calls it, perceives onely those images and does not perceive the external Object immediately but by the Medium of the images in the Brain. This Notion of our perceiving external Objects, not immediately but by certain images or Species of them conveyd to the Mind by the external Senses, seems to be the most ancient philosophical Hypothesis we have upon the Subject of perception and to have with some Small Variations, retained its authority to this Day. Aristotle, as was before observed, maintained that the Species, forms, or images of external Objects coming from the Objects themselves enter by the Senses & are impressed upon the intellect. The Epicureans held the same thing with regard to slender films of subtile Matter coming from the Object, as Aristotle did with regard to his immaterial forms or Species. Aristotle & his followers thought that all the objects of human Understanding ente⟨r⟩ by the Senses, & that it is by the internal Powers of the Mind refining & spiritualizing the Notions we get by the Senses are by the internal powers of the Mind refined & Spiritualised so as at last to become the Objects of Science. Plato on the other hand had a mean opinion

[2v] of all | the knowledge we can get by the Senses. He thought it did not deserve the Name of Knowledge & could not be the foundation of Science because the Objects of Sense are in a perpetual fluctuation. All Science according to him must be employed about those eternal & Imutable Ideas which existed before the Objects of Sense, & Are allways invariably the Same. In this there was an Essential Difference between the Systems of those two Philosophers. The Notion of Eternal & Imutable Ideas which Plato borrowed from the Pythagorean School was totally rejected by Aristotle, & it was a Maxim of his that there is nothing in the Intellect which was not first in the Senses.

But notwithstanding this great difference in those two ancient Systems, they might both agree as to the Manner in which we perceive Objects by our Senses. And that they did so I think is probable because Aristotle as far as I know neither takes Notice of any difference between himself & his Master upon this point, nor lays claim to the Invention of his Theory with regard to the Manner in which we perceive external Objects. It is still more probable from the hints which Plato gives in the 7 Book of his Republick of the Manner in which we perceive Objects by Our Senses. Which he resembles to Persons in a deep & dark Cave, seeing, not the things themselves that are without but onely their Shaddows by a small Light let into the Cave

It seems probable therefore that the Pythagoreans & Platonists agreed with Aristotle & the Peripateticks in this general Theory of Perception, to wit that external Objects of Sense are perceived onely by certain images or shaddows of them let in to the Mind or into its presence chamber as into a *Camera obscura*

The Notions of the Ancients, were very various with regard to the Seat of the Soul. Since by the improvement of Anatomy, it has been discovered that the Nerves are the Instruments of Perception & of the Sensations that accompany it, and that the Nerves do all alternately terminate in the Brain; the Opinion has generally prevailed among Philosophers that the Brain is the Seat of the Soul. And that she perceives nothing immediately but the images that are brought there. Des Cartes observing that the pineal Gland is the onely part of the Brain that is single, & that all the other parts are double, & thinking it improbable that the Soul should have two habitations was determined by this to make that Gland the Souls habitation, to which by the Animal Spirits intelligence is brought of all Objects that affect the Senses. |

[1r] As to the first point That the Soul has its Seat in the Brain. This surely is not so well established as that we can safely build other principles upon it. There have been various Opinions and Much Disputation about the place of Spirits, whether they have a place or not & if they have how they Occupy that place; but after Men had fought in the Dark for Ages about these points, the wiser part seem to have left of Disputing about them, as Matters beyond the reach of the human faculties.

We consider the Body without the Soul as a Hydraulick Machine made up of a prodigious variety of parts most admirably fitted for various purposes. But a Power is wanting to begin and to continue the Motions & produce the effects for which it is fitted. This Power we call the Soul. But where shall we place it? If we confine it to the Brain then it will follow that the Brain is the onely Part where all motions & Effects produced by the Vis Vitæ begin, & that they are carried on through the other parts by mere mechanism The Power from the brain that makes the Heart and arteries to beat the Muscles to contract and Relax with so great force The power that makes the Stomach Digest the Lungs to play & the power that produces nourishment and Growth & that preserves the whole from that putrefaction which would soon be produced by the Air and circumambient bodies It belongs ⟨to⟩ anatomists and Physiologists to determine whether this be so or not. Their curious and valuable discoveries have not hitherto tended to confirm this Hypothesis but rather to the contrary. The first Motion that appears in Animals is the pulsation of the heart.

Nihil aliud est imaginari quam rei corpor⟨e⟩æ figuram seu imaginem contemplari Meditat 2.

per Solum intellectum, percipio tantum Ideas de quibus judicium ferre possum Med 4. p 27

Neque opus est me in utramque parte⟨m⟩ ferri posse ut sim liber; sed contra quo magis in unam propendeo, quia rationem veri et boni in ea evidenter intelligo tanto liberius eam eligo Med 4. 28

Satis est quod possim unam rem absque altera clare et distincte intelligere, ut certus sim unam ab altera esse diversam, quia potest, saltem a Deo, seorsum poni.

IV

3061/12, 2v   Disquisitions relating to matter & spirit J Priestly 1777

Newtons first rule of philosophizing misquoted, an essential part of it being omitted p 2

The vulgar opinion he makes to be that matter is a solid & inert Substance

Resistance on which alone our opinion concerning the solidity or impenetrability of matter is founded, is never occasioned by solid matter, but by a power of repulsion. It will also appear that without a power of attraction there cannot be any such thing as matter. This power is in reality absolutely essential to its very nature & being. For when we suppose bodies to be divested of it they come to be nothing at all.

A ray of Light passes through a transparent body without a single particle being reflected till it comes to the opposite side.

V

3061/13, 1r   Free discussion.

p. 10 Price. Matter if it be anything at all, must consist of solid particles or Atoms occupying a certain portion of space, and therefore extended, but at the same time simple and uncompounded, and incapable of being resolved into any other smaller particles; and it must be the different form of these primary particles and their different combinations & arrangement that constitute the different bodies and kinds of Matter in the Universe. This seems to have been Sir I Newtons Idea of Matter.

149 Priestly. To guard against all mistake, it may be more advisable that in treating the subject philosophically, those words (accountableness, praise and blame, merit and demerit) be disused.

296 Priestly There does not appear to me to be, in the whole compass of reasoning, that I am acquainted with, a more conclusive Argument, than that for the doctrine of Necessity from the consideration of the Nature of Cause & Effect.

293 Id. In the case of the beam (suspending equal weights at both ends) it is immediately perceived that bearing an equal Relation to both the weights, it cannot possibly favour one of them

more than the other; and it is simply on account of its bearing an equal Relation to them both that it cannot do this.

ibid The Mind itself can no more be the cause of its own determination, than the beam of a ballance can be the cause of its own inclination.

249 One single Center could not be said to be divisible, or even to occupy any portion of space

366. It is as evident that the Brain thinks as that the Magnet attracts Iron.

367 I consider Man as preserving his Individuality or Identity, in the same manner as a tree does

381 The decisions of common Sense, as far as they go are uniformly on the side of the Doctrine of Necessity

390 By a Mans *making his own fortune* I mean that his success depends upon his actions, as those depend upon his volitions, and his volitions upon the motives presented to him. He will therefore necessarily be roused in proportion to the strength of his propensity (to happiness) and his belief of the necessary connection between his end and his endeavours, and nothing but such an opinion as that of philosophical Liberty, which destroys that necessary connection, can possibly slacken his endeavours.

400 The Cause of a Being or Substance must be a Being or Substance also

Disquisitions page 125. It is true that a property, such as I conceive the power of thinking to be, cannot exist without its Substance. |

[1v]   Disquisitions pag 235 It is evident that Brutes have the Rudiments of all our Faculties without Exception; so that they differ from us in degree onely, and not in kind

237 But he (Cudworth) injures the brutes very much when – he supposes them to be destitute of Morality and Liberty

Introductory Essays to Hartleys Theory of the human Mind pag 27 "And having attained to the knowledge of *general Truths*, the Idea or Feeling which accompanies the Perception of Truth, is transferred, by association to all the particulars which are comprized under it, and to other propositions that are analogous to it; having found by experience that when we have formed such conclusions, we have not been deceived"

Part Three: Materialism

Preface to the 3d Vol of Institutes of Natural & Revealed Religion London 1774

p. 12 And if this (to wit the Non existence of an external World) appears to any person a more Natural & simple Hypothesis, to account for our Ideas, and therefore preferable to the Supposition of a real External World – I leave him to his Imagination from which no evil that I know will result

Examination p 321. But this Philosopher Dr Price had more good sense than to load his Scheme with the belief of the real Existence of the external World.

Examination. Introductory Observations p 57. It may occur to some persons, that, since we are not properly *conscious*, or *know in the first instance*, any thing more than what passes within ourselves, that is, our own sensations & ideas, these may be impressed upon the Mind without the help of any thing external to us, by the immediate agency of the Author of our Being. This no Philosopher will say is impossible, but, of two hypotheses to account for the same Phenomenon, he will consider which is the more probable, as being more consonant to the course of Nature in other respects

Ibid p 60 It is not true that we necessarily believe the existence of external Objects *as distinct from our Ideas of them*. Originally we have no knowledge of any such thing as Ideas, any more than we have of the Images of Objects on the Retina; & the moment we have attained to the knowledge of Ideas, the external world is nothing more than an hypothesis to account for those Ideas, so probable indeed that few persons doubt of its real Existence & of its being the cause of our Ideas.

Ibid p 155 Our Author had nothing in View but this same innocent theory of Berkeley

Ibid p 181. In short I have no conception that the Man whose Mind is capable of entertaining and duly contemplating what is called the doctrine of Necessity as | unfolded by Dr Hartley, can be a bad Man, nay that he can be other than an extraordinary good one.

Disq p 207 That the terms Essence Substance &c &⟨c⟩ are merely a convenience in speech

Disq Introd p 38 Matter is not that inert Substance that it has been supposed to be. I define it to be a Substance possessed of the property of Extension & of powers of Attraction & Repulsion Disq pag 72. But did not this Writer (Letters on Materialism) know that it is even demonstrable that Matter is infinitely divisible

VI

Materialism & Fatalism are very ancient Opinions and have been so much canvassed in every Age since Philosophy had an Existence among Men, that it might be thought that nothing could be said for or against them in our Day which has not been said in former Ages.

Both have been thought unfriendly to Religion and Morals; and therefore have generally been defended by those who wished to undermine, and opposed by those who wished to support those great Interests of Mankind.

Yet we have lived to see a Philosopher & Divine of great Celebrity, who has not onely endeavoured to give new Light to these ancient opinions; but to bring them over to the side of Religion.

The Revd Dr Priestley has taught us that Materialism & Fatalism, together with a third associate, to wit Socinianism, are the three great Pillars on which Religion & Morality ought to be founded.

It is onely the first of these Pillars, Materialism, we have to do with at present, which that Philosopher, from his zeal for Religion, without neglecting the trite Arguments urged in its Defence from the time of Lucretius down to our age, has endeavoured to support by new Arguments, drawn from Principles of modern Philosophy, by the Rules of Philosophising laid down in the *Principia* of Sir Isaac Newton.

The Sum of the System is, That Man is not, as is commonly believed, compounded of two Substances, to wit | an unthinking Substance which we call the Body, and a thinking Substance which we call the Mind, but is wholly Material, That the thinking part of Man is his Brain, which requires onely a proper Organization to produce Sensation, Thought, Reasoning, and all the Mental Powers of Man.

So far this System is common to Dr Priestley and former Materialists. But the addition made by him to the ancient System is, That Matter is not an Inert Solid & Impenetrable Substance, as it has commonly been supposed to be; It is neither Solid & Impenetrable, nor is it Inert, & in reality it has no properties but Extension, & inherent powers of attraction & repulsion.

Part Three: Materialism

Thus he rejects two properties which were conceived to be essential to Matter, & which indeed made most people suspect that it is not a very proper kind of stuff to make a human Soul; these are Solidity or Impenetrability, and Inertia or Inactivity. In place of these he has discovered a new essential property of Matter, namely inherent powers of Attraction & Repulsion.

That there are in Nature Powers of Attraction & Repulsion is known and acknowledged by all Philosophers, but they were conceived not to be inherent in Matter but forces impressed upon it by some unknown Cause. This was the Opinion of Sir Isaac Newton the discoveror of those Powers. But this Philosopher has found them to be inherent in Matter and essential to it; and thereby hopes to remove the prejudice against Materialism arising from the conceived Solidity and Inertia of Matter.

Of this new Idea of Matter I take this Philosopher to be the Inventor; and he supports it by some discoveries of Modern Philosophy, & chiefly by Sir Isaac Newtons rules | of Philosophizing.

It is not my Intention to inquire whether this new Idea of Matter makes it more fit to be the subject of thought than that which was entertained before, or whether Extension & powers of Attraction and repulsion (the onely properties which Dr Priestley will have to belong to Matter) may be so modified and mixed as to produce Sensation Memory Volition Reasoning, & all the Faculties & Powers of a human Soul. This would onely lead to the same train of reasoning which has often been used against the ancient System of Materialism, which I shall leave to stand or fall by its own Merit. The Observations I proposed to make, relate onely to the new Idea of Matter given by Dr Priestley, and to the aid he borrows from modern Philosophy & chiefly from Newtons Rules of philosophizing in support of it.

On this Subject I have had the honour to read two Discourses to the Society, & am now to finish what I intend upon it.

I observed First That there are some Opinions of very eminent modern Philosophers, which are subversive of Materialism, which therefore we might have expected Dr Priestly would have refuted in order to lay a foundation for his System. But we are so far disappointed that he seems to have adopted them, or thought more favourably of them than became a zealous Materialist.

The first I mentioned was That we have no Idea of Substance,

that it is a thing invented by Philosophers to support Accidents, as
the Indian Philosopher invented his Elephant to support the Earth.
This was maintained by | Locke Berkeley & Hume; and I mentioned
several passages of Dr Priestley by which he appears to give his
suffrage to this opinion. If this be so, to dispute whether Man be
one Substance or compounded of two, must be a dispute merely
about a word of which we have no Idea; a dispute about the unity
or duplicity of an imaginary thing.

A second Opinion subversive of Materialism is that of Bishop
Berkely. That there is no such thing as Matter in the Universe.
This Dr Priestley thinks an innocent hypothesis which will do no
Man harm. He commends Dr Price for having more good Sense
than to load his System with the belief of the existence of an external
world. In other places however he says the existence of Matter is a
hypothesis so probable that few persons doubt of it, & in two
different works, & in the very same words, has given a detail of
this probability. I endeavoured to shew that this probability on
which this Philosopher rests the existence of Matter & consequently
the foundation of Materialism takes for granted the thing to be
proved & has no force at all against Bishop Berkeleys System.
Unless therefore a more solid Refutation of Berkeleys System be
found than he has given there is no place left for Materialism.

A Third opinion on which the two former are grounded and
which therefore is subversive of Materialism, is that fundamental
Principle of Mr Lockes System That all our Ideas are what he calls
Ideas of Sensation or Ideas of Reflection. From this principle it
was that Locke Berkely and Hume by just inference concluded
that we have no Idea of Substance & to this conclusion Dr Priestley
assents. From this Principle Bishop Berkeley by invincible reasoning
concluded that there is no Matter in the Universe. This Conclusion
was confirmed by Mr Hume and Dr Priestley finds nothing to oppose
to it, but what he accounts a probability, but what I endeavoured
to shew is in reality mere Sophism. Yet to this Principle of Mr
Locke so unfriendly to Materialism Dr Priestley most firmly ad-
heres | He conceives it to be the cornerstone of all rational know-
ledge of our selves. His Zeal is k⟨i⟩ndled against a set of *pretended*
Philosophers who have called it in Question of whom he says the
most conspicuous and assuming is Dr Reid professor of Moral Philosophy in the University of Glasgow. He acknowledges that the

hypothesis of there being no external World is by no means so shocking to his Understanding as the supposition that we really perceive things that are external to us. Yet if this cornerstone of Mr Lockes System leave us no Idea of Substance; if it leave us no material World, Materialism can find no foundation till it be removed.

I considered next the Account he gives of Sir Isaac Newtons Rules of philosophizing, to which he professes an uniform and rigorous adherence, & requires that his reasoning be tried by this Test onely & by no other. After this profession he tells us what the first and what the second of these Rules is as laid down by Sir Isaac Newton. The third he never mentions, though it touches his subject as nearly as either or both the other two. And as to the first and second, which he professes to give as laid down by Sir Isaac Newton, of them he has given us a lame & erroneous translation, by which they are as well fitted to support the occult Qualities of Aristotle or the Vortices of Des Cartes or ⟨the System⟩ of Dr Priestley as that of Sir Isaac Newton. This I endeavoured to make apparent in the first Discourse.

In the second Discourse I considered what Dr Priestley has said to overturn those two Qualities of Matter, which modern Philosophers since the time of Galileo, & especially since that of Newton, have believed to be inherent in it, to wit Solidity or impenetrability, & Inertia or Inactivity. A Recapitulation would make this Discourse too tedious, | but upon the whole I could find nothing to overturn those qualities, or to weaken the evidence on which they have been believed to be inherent in Matter.

It is not strange that a Materialist should be averse to acknowledge the Inertia of Matter; but that denying the Inertia of Matter he should profess to build his System upon the three Rules of Philosophizing laid down by Newton, is indeed very strange, &, I think, can be accounted for onely by this, that he misunderstood the two first of those Rules, and overlooked the third.

But perhaps this Philosopher may be more happy in building than in overturning. I proceed therefore to consider, whether he has given a firm Foundation to that Quality of Matter which he has discovered, namely, That it has inherent powers of Attraction and Repulsion.

That there are in Nature various powers of Attraction and

Repulsion, by which the parts of Matter are affected, was discovered by Sir Isaac Newton, and is acknowledged by all modern Philosophers. And the Question between Dr Priestley on the one hand & Sir Isaac Newton on the other, is not about the Existence of those Powers but about their Subject, whether they be inherent in Matter, or forces impressed upon it by some forreign Cause, so that in consequence of those powers Matter does not act, but is acted upon, does not attract & repell but is attracted and repelled.

It is one thing to say &c

VII

May 1788 Read Disquisitions Relating to Matter & Spirit by J Priestly Lond 1777

Pref pag 15 the Author became satisfied that if we suffer ourselves to be guided in our inquiries by the universally acknowledged *Rules of Philosophizing* we shall find ourselves intirely unauthorized to admit any thing in Man besides that Body which is the object of our Senses

Sect 1. Of the Nature & essential Properties of Matter.

The Author, with Sorrow, calls upon his Reader to recur to the Universally received *Rules of Philosophizing*, laid down by Newton He professes an uniform & rigorous adherence to them, & requires that his own Reasoning be tried by this & by no other Test. It is strange that after this solemn Parade he should give us the Rules he alludes to mangled and misinterpreted. "The first of these Rules as laid down by Sir Isaac Newton, is that we are to *admit no more Causes of things than are sufficient to explain Appearances*; and the second is that, *to the same Effects we must, as far as possible, assign the same causes.*"

Sir Isaac Newtons first Rule is That no more Causes of Natural Things ought to be admitted than such as are both true, and sufficient to explain their Appearances. The first of these Conditions in Sir Isaac Newton's Rule is required as no less necessary than the second but Dr Priestley's Rule requires the second onely without the least mention of the first. The Vortices of Des Cartes & the Vibrations & Vibratiuncles of Dr Hartley, according to Dr Priestleys

Rule, ought ⟨to be⟩ admitted as Causes of Natural Things for they explain Appearances in a very ingenious manner; but they are excluded by Sir Isaac Newtons, because they want the essential Condition of Truth. We have no evidence of their Existence To interpret this capital Rule of Sir Isaac Newton as if it gave a Sanction to Theories which have no other evidence of their Truth but that of explaining Appearances, is to contradict its main design.

Sir Isaac Newtons second Rule is Of Natural Effects *of the same kind* the Causes are the Same. According to Dr Priestly it is that, *to the same Effects we must, as far as possible, assign the same Causes*. Whether the Rule be mended or made more distinct by putting Effects *in general* in place of *natural* Effects, & and the *same* Effects in place of Effects *of the same kind* is not worth while to consider Such accurate Writers as Sir Isaac Newton are injured, but not enlightened, by a loose paraphrase.

Sir Isaac Newton does not direct us to apply this Rule *as far as possible* as if the danger were of not applying it when it ought to be applied this would not have been agreeable to the Spirit of his Philosophy. When Natural Effects are evidently of the same kind, the conviction which all Men have of the Uniformity of Nature is a sufficient inducement to impute them to the same Cause, & they need not to be required to do this as far as possible. The danger lies on the other side, lest our proneness to assign similar Effects to the same Cause should lead us to take Effects to be of the same kind & therefore to have the same Cause because they have some resemblance. To obviate this danger Sir Isaac Newton illustrates the Rule onely by Examples of those Effects that are Evidently of the same kind such as Respiration in Men & Brutes, & the descent of heavy Bodies in Europe and in America. Had he thought fit to

[2v, margin] enlarge upon the | application of his Rule, there is Reason to think, that he would rather have warned Men that it is to be applied with great Caution and Prudence than directed to apply it as far as possible because most of the false Theories in Philosophy have been owing to the misapplication of it. |

[1r] What the Author endeavours to prove in this Section is that Matter is not solid, impenetrable & endowed with a Vis Inertiæ. By this he would overturn Sir I. Newtons Laws of Motion by his Laws of Philosophizing. He seems to think that Solidity when spoke of as a Property of all Matter means onely hardness

& therefore requires a certain mutual Attraction of the Parts. That if this Attraction be removed all solidity must be removed with it, & that the body will be annihilated. Because it cannot be proved that bodies by impulse do really touch, he thinks it follows that the Resistance of Bodies is solely caused by repulsion; not | considering that there could be no repulsion without resistance and that the resistance of a Body to Motion is the same whether it be moved by a repelling power acting at some distance or by contact, The resistance to Motion is as the Quantity of Matter to be Moved & the Quantity of Motion produced, and not as the repelling power. The weakness of this reasoning is the same as if one should maintain that I do not raise a weight because I do not touch it but raise it by a pully or a lever.

Both Solidity & impenetrability he conceives to depend on hardness and consequently upon attraction of the Parts. "It will perhaps be said, says he page 6, that the particles of which any solid Atom consists, may be conceived to be placed close together without attraction. But then this Atom will be entirely destitute of compactness, & hardness which is requisite to its being impenetrable."

The powers of Attraction and Repulsion he makes essential to the being of Matter, but he does not suppose that they are *selfexistent in it*. I know not the meaning of selfexistence here. Yet he thinks it agreable to the Rules of philosophizing to consider all the constant Effects of any Substance as produced by powers properly belonging to that Substance, whether they be necessarily inherent in it, or communicated to it pag 9.

Page 18 "The onely Reason why the principle of Thought or Sensation has been imagined to be incompatible with Matter goes upon the Supposition of impenetrability being the essential Property of it, & consequently that solid extent is the foundation of all the properties it can possibly sustain"

To conclude what we have to say upon these two Sections, That we have no certainty that in any of the Operations of Nature or in any casual concourse Bodies come into actual Contact was long since observed by Newton, and the first hint of it was sufficient to convince all Men who understand his Philosophy that on account of the imperfection of our Senses It is impossible we should have evidence of such Contact. It has likewise been long believed by all who know the Principles of Natural Philosophy, that no Body, nor even the Elementary Particles of Matter can have any degree of Hardness without some attraction or Power of Cohesion, which Newton & those who follow him conceive not to be essential to Matter but to be a force impressed upon it. These two Principles Dr Priestly might very well have assumed without any Laboured proof. Nor is there any Need of Newtons Laws of Philosophizing to confirm things in themselves so evident, & which I apprehend no person denies who has any knowledge in Natural Philosophy.

What really needed proof is the consequences which Dr Priestly draws from these Principles. to wit that Matter is neither solid nor impenetreble nor has a *vis inertiæ* |

It were to be wished that Dr Priestly had explained what he means by solidity, what by impenetrability, & what by the vis Inertiæ

Solidity in common Language signifies a quality which does not belong to all Bodies but to some onely, & therefore is opposed to fluidity, which is the property of Bodies that are not solid.

But when Solidity is mentioned by Philosophers as a property of all Bodies, it must have another meaning; and then it means that every body occupies its place in such a manner as to exclude other Bodies from occupying the same place at the same Time.

If the Attraction & Repulsion which we observe between Bodies be really an active Power inherent in these Bodies, & not a force impressed upon them by some external Cause It will follow first that Bodies act where they are not, which I take to be an absurdity. Secondly this inherent Power of acting without any Motive, and without any external impulse seems according to Dr Priestleys System to be impossible, & to be what he calls an Effect without a Cause. If it be true, as he says, Free discussion page 293, *That the Mind can no more be the Cause of its own determinations than the beam of a ballance can be the cause of its own Inclination*, are we to think that a Body is the Cause of its Attractions & Repulsions.

VIII

11 Junii 1788 Philosophiæ Recentioris a Benedicto Stay Raguisina in Rom: Archigym: Pub Eloquentiæ Profess: versibus Traditæ Lib 10 ad Sylvium Valentium Card. cum adnotationibus & Supplementis P. Rogerii Josephi Boscovich. S.

Jesu in Collegio Rom. Pub. Matheseos Profess. Tom 1 Romæ 1755

pag 27 Not. Ego quidem arbitror corpus constare punctis prorsus inextensis a se distantibus

pag 57. In Newtons first Law of Philosophising, both Stay & Boscovitch omit an important part. plures causas admitti non debere quam quæ *et veræ sint*

63 Not In mea quidem Sententia de Materia composita punctis indivisibilibus, inextensis &c

Suppl 337 Of the Qualities good or bad of things created & consequently limited there is not any quantity that can be called absolutely the greatest, the best or the worst the greatest possible Number, the greatest possible Extent, the greatest possible quantity of happiness of Virtue in finite things is a Contradiction, because it is the Nature of such things to be capable of increase without any limit. This observation Boscovitch very ingeniously applies in answer to the Leibnitzian Theory that the Deity of all possible Worlds must have chosen that which was best. With Regard to the Harmonie Preetablie he observes very Justly that it leaves no Evidence of the Existence of things External to us.

343 Space & Time considered abstractly are not realities but possibilities and therefore may without error be said to be eternal infinite & uncreated.

351 Boscovitch maintains that it is impossible to distinguish real from absolute Motion, & replies to what Newton says on that point.

355 He observes very justly that in his Hypothesis of the Nature of Matter there is no limit to its density or Rarity whereas in the Common Hypothesis there is an absolute density incapable of increase

363 &c The Inertia of Matter cannot be proved a priori. As of the Motions we observe we have no certain Means of discerning which are real & absolute & which are relative, so the vis Inertiæ of Matter which we deduce with great probability from Experience is onely a relative Inertia. The whole Newtonian System is relatively just & true yet it is possible without any contradiction to it that the Earth may absolutely be at rest.

371 The Cartesians who maintain that matter is not the physical or Efficient Cause but onely the Occasional Cause of the Effects

on other matter which we ascribe to it can have no solid Objection to the Law of Gravitation, as Matter may be the occasional Cause of any change in other Matter, at a distance as well as in Contact.

IX

Some Observations On the Modern System of Materialism.

Introduction.

By the Modern System of Materialism I mean that which is advanced by the Reverend Dr. Priestley in his *Disquisitions relating to Matter and Spirit* published Anno 1777, and defended in his *Free Discussion of the Doctrines of Materialism and Philosophical Necessity*, and *Illustrations of the Disquisitions* published Anno 1778.

As I know no other Author who has explained and defended this System, I take the principles of it from him only.

The sum of them is, That Man is not, as is commonly believed, compounded of two Substances, to wit, an unthinking Substance which we call the body, and thinking Substance which we call the Mind, but is wholly Material; That Matter, however, is not that inert, solid and impenetrable Substance which it has commonly been supposed to be; That, in fact, it has no properties but those of extension and inherent powers of Attraction and Repulsion.

This account of the properties and powers of Matter I take to be new, and that which distinguishes Dr. Priestley's System from other Systems of Materialism, and it is to this part of his System that the following observations chiefly relate.

Chap. I. Of the connection of this System with other Philosophical Opinions.

There are some Philosophical Opinions supported by | very eminent modern Philosophers, which if they be true, subvert the very foundation of Materialism, which therefore we might expect would have been refuted and removed in order to build this System. Dr. Priestley, however, far from refuting those opinions, speaks more favourably of them than a consistent and zealous Materialist ought to do.

I. There are two contradictory opinions with regard to what we call a Substance as distinguished from the things we call Qualities.

The ancient Philosophers thought it selfevident that the things we call Qualities cannot possibly exist without a Substance in which they are inherent, to which they properly belong, and from which they necessarily result, as the properties of a mathematical figure result from its nature and definition. They conceived therefore that figure cannot, without absurdity, be supposed to exist but in something figured, nor motion but in something moved, nor Judgment and Memory but in something that judges and remembers.

This has been the opinion of the Vulgar in all ages, and is implied in the grammatical structure of all languages, and I cannot help assenting to it.

It is true indeed that we have no immediate perception of Substance either by sense or by consciousness, but we are led by our constitution to infer its existence from the qualites which we immediately perceive, and though the relation between a Substance and its qualities be in some respects obscure, it is easily distinguished from all other relations.

If this vulgar opinion concerning Substance be true, there is no impropriety in enquiring whether Soul and Body be two distinct Substances, or one and the same.

But the modern Philosophy, introduced by Des Cartes and improved by Locke, Berkeley, and Hume, is inimical to | this notion of Substance, and leads us to conceive it as a vulgar prejudice.

Des Cartes the Father of this System seems to have had some apprehension of this, by his making extension to be, not a quality of Matter, but the very essence of it, and Thought, not to be an operation of Mind, but its essence. The different modes of thinking, such as judging, willing, remembering, his followers chose to call *Modifications of the Mind*, rather than *Operations of the Mind*, the first of these phrases implying that the Mind is nothing but Thought which may be variously modified, the last implying that the Mind is not Thought, but a Being or Substance which has the power of thinking.

Mr. Locke saw more clearly that his System did not admit any idea of Substance. He says expressly, "That we neither have nor can have any idea of Substance either by Sensation or reflection", and "that we signify nothing by that word but an uncertain supposi-

Part Three: Materialism

tion of, we know not what idea, which we take to be the *Substratum* or support of those ideas we do know." *Essay 1.4.18*. He observes, "That they who first ran into the notion of *Accidents*, as a sort of real beings, that needed something to inhere in, were forced to find out the word *Substance* to support them. That had the poor Indian Philosopher (who imagined that the earth also wanted something to bear it up) but thought of this word *Substan⟨c⟩e*, he needed not to have been at the trouble to find an Elephant to support it, and a Tortoise to support his Elephant: The word *Substance* would have done it effectually." *Essay 2.13.19*.

It is well known that Bishop Berkeley and Mr. Hume have employed their reason and wit against this poor word Substance, and treated a *Substratum* of qualities as a Chimera.

[4] Now, this modern opinion concerning Substance seems to me in|consistent with the belief that the Soul is either a material or an immaterial Substance. Mr. Hume is perfectly consistent with himself when he makes Body to be only a certain collection of Qualities to which we give one name, and Mind to be a certain succession of Ideas and Impressions.

If I believe that Substance is a word without a meaning, or that it is an imaginary thing, invented to support Accidents, as the Indian invented his Elephant to support the Earth, how can I seriously believe that either Body or Soul is a Substance? Can it be a serious Question, worthy of a laboured discussion, whether Man be one Substance or compounded of two, if the existence of Substance be imaginary or uncertain?

It seems evident therefore that one who holds the System of Materialism to be true and important, ought not to be dubious with regard to the existence of Substance. Yet Dr. Priestley seems more dubious with regard to the last of these points, than accords with his zeal for the first.

He says, indeed, *Disquisitions page 125*, "It is true that a property, such as I conceive the power of thinking to be, cannot exist without its Substance." This is perfectly consistent with Materialism. For if neither properties nor powers can exist without a Substance to which they belong, it follows necessarily, that the body which has properties must be a Substance, and that the Mind which has powers must be a Substance, and it may reasonably be enquired whether they be one and the same Substance or not.

In many other places, however, this Author seems to be very dubious whether there be any such thing as Substance, and very positive that we have no proper idea of it.

Disquisitions page 104. "In fact we have no proper idea of any essence whatsoever". *Ibid.* "The term *Substance* or Essence therefore is, in fact, nothing more than a help to expression, as we may | say, but not at all to conception".

If the term Substance express any conception of the mind, though even an obscure conception, it must be a help to conception; and if it be no help to conception, it can express no conception clear or obscure, and then I apprehend it can be no help to expression but an incumbrance upon it, as all unmeaning words are.

Disquisitions page 107. "The properties or powers being different, the Substance or Essence (if it be any convenience to us to use such terms at all) must be different also".

Here, I would be glad to know how, without the use of such terms, we should express the fundamental principle of Materialism, That Soul and Body are not two different Substances, but one and the same.

Free Discussion page 364. "Having again and again observed, what I believe will be universally admitted, that by the words *Being*, *Substance* or *Thing*, we only mean the unknown, and, perhaps, imaginary support of properties."

If this be so, will it not follow, that the whole dispute about Materialism, is a dispute about the unity or duplicity of an unknown, and, perhaps, an imaginary thing.

It seems therefore incumbent on those who support the doctrine of Materialism as a doctrine of real importance in Philosophy and in Religion, not only, not to adopt the scruples that have been raised by modern Philosophers about the existence of Substance, but to shew that they have no solid foundation, and that however imperfect or obscure our notion of Substance may be, we must admit their existence, and that qualities cannot subsist without them.

II. Another opinion subversive of Materialism is that which denies the existence of Matter.

If one believe, with Bishop Berkeley, that there is no | such thing as a material world, that ideas and immaterial thinking Beings are the only existences in the Universe, it is impossible that he can

Part Three: Materialism 177

believe at the same time that the human Soul is a piece of Matter. Bishop Berkeley thought it no small recommendation of his opinion, that, if it be true, Materialism is effectually refuted. And, indeed, the contradiction between these two opinions is selfevident. Every argument for Materialism supposes the existence of Matter, and, if matter do not exist, must be fallacious and inconclusive.

If the existence of Matter be an hypothesis only, brought to account for the production of ideas in our Minds, it will follow, that all that can be said for Materialism, is grounded upon an hypothesis; and this, to those who have a just taste in Philosophy, must appear to be a very slippery foundation.

It might therefore be expected of Materialists, that in order to lay a solid foundation for their system, they should shew that the subtile arguments of Bishop Berkeley against the existence, and even against the possibility of a material world, have no force.

Dr. Priestley expresses a more favourable opinion of Bishop Berkeley's system, than a confirmed Materialist ought to entertain. *Institutes of Natural and Revealed Religion, Pref page 12.* "If this appears to any person a more natural and simple hypothesis to account for our Ideas, and therefore preferable to the supposition of a real external world; by means of which and of a more general agency of the Deity, the same ideas may be presented to thousands and millions of Minds, I leave him to his imagination, from which no evil that I know will result."

Surely this evil will result, that he cannot be a Materialist. *Examination of Reid, Beattie and Oswald, Introd page 60.* "The moment we have attained to the knowledge of Ideas, the external world is nothing more than an hypothesis, to account for those ideas, so probable indeed, that few persons seriously doubt of its real existence, and of its being the cause of our Ideas." | *Ibid. page 155*, He calls it, "This same innocent theory of Berkeleys" And *page 312*, He says of Dr. Price, "This Philosopher had more good sense than to load his scheme with the belief of the real existence of the external world."

It happens that the scheme of Materialism is necessarily loaded with the belief of the real existence of the external world; and therefore it is incumbent upon the defenders of it to support that load.

It is true, Dr. Priestley thinks the existence of the external world a probable hypothesis, because it accounts for appearances. Supposing it to be so, the system of Materialism can have no more probability than that of an hypothesis that accounts for appearances. And those who have a just taste in Philosophy, pay no regard to such hypotheses.

Sir Isaac Newton has told us, that they have no place in Philosophy. "Hypotheses non fingo. Quiquid enim ex phaenom⟨e⟩nis non deducitur, hypothesis vocanda est, et hypotheses, seu Metaphysicae, seu physicae, seu qualitatum occultarum, seu Mechani-⟨c⟩ae, in philosophia experimentali locum non habent." And his great Friend Roger Cotes, describing the experimental method of philosophizing, says, "Nihil autem principii loco assumunt quod nondum ex phaenomenis comprobatum fuerit. Hypotheses non comm⟨i⟩niscuntur, neque in Physicam recipiunt, nisi ut quaestiones de quarum veritate disputetur."

The probability of this hypothesis Dr. Priestley places in this. "That it exhibits particular appearances as arising from more general laws." This is a probability common to thousands of hypotheses which time has discovered to be false. But what is its claim to this probability?

Dr. Priestley has in two different works, and in the same words, shewn this at full length, first in the preface to the third volume of his *Institutes, pages 12 and 13*, and afterwards in his introductory observations to his *Examination of Reid, Beattie and Oswald, pages 58 and 59*. The words are, "Half the inhabitants of the globe, for in|stance, may be looking towards the heavens at the same time, and all their minds are impressed in the same manner: All see the moon, stars and planets precisely in the same situations; and even the observations of those who use Telescopes correspond with the utmost exactness. To explain this, Bishop Berkeley says, that the divine Being, attending particularly to each individual mind, impresses their *Sensoriums* in the same or in a corresponding manner, without the medium of any thing external to them. On the other hand, I, without pretending that his scheme is impossible, where divine power is concerned, think, however, that it is more natural to suppose, that there are really such bodies as the moon, stars and planets, placed at certain distances from us, and moving in certain directions, by means of which, without such an agency as he supposes, all our minds are necessarily impressed in this corresponding manner."

This is the probability on which the existence of an external world, and, of consequence, the truth of the system of Materialism is grounded by Dr. Priestley; and he acknowledges, that the hypothesis of there being no external world, is by no means so shocking to his understanding, as the supposition that we really perceive things that are external to us, and do not judge of all things that are without ourselves, by notices perceived within, how mistaken soever we may be in our judgments concerning them.

Were I to examine this probability on the principle which Berkeley assumes, and which to Dr. Priestley's understanding appears more evident than even the existence of an external world, I would say, that an inhabited globe, its numerous inhabitants, their observations and telescopes, as well as the Moon, stars and planets which they observe, are, according to this principle, nothing but Ideas in my Mind. The external existence of things corresponding or similar to these Ideas, is the very thing to be proved; and, in this argument, that external existence is taken for granted. So that though the consequence were not only probable but demonstrative, the argument is only what | Logicians call a *Petitio principii*.

Besides this, we have no reason to think that an external world could possibly enter into the conception of a Being who never perceived any thing external. Our conceptions are limited by our perceptions. A man who never saw colours cannot conceive colours, nor can one who never heard sound conceive sound. We can in imagination mix and compound the materials got by perception in various ways, but to add other Materials to them is no more in our power than to create. According to Dr. Priestley, all that we really perceive is only sensations in our own Minds. No mixture or composition of Sensations can make up the conception of an external world. Sensations are fleeting and transient things, like the images in a Mirror; the external world has a real and permanent existence. Sensations cannot be but in a sentient mind, as their subject, the external world, has an existence given it independent of any subject, and is totally of a different kind from any sensation or composition of Sensations.

Let us suppose that man were such a Being as Dr. Priestley conceives him to be, that is, a being who really perceives nothing external to himself, and judges of all things only by notices perceived in his own mind; and let us suppose that in consequence of his reflection upon his inward notices, he puts this question to himself, What is the cause of his ideas and sensations, and of that regularity with which they are produced, with manifest marks of wisdom superior to his own, and of benevolent design? How would he judge in this case?

Would he attribute these effects to an unthinking cause such as a material world, granting that he could form a conception of such a cause? Surely, if he had understanding he would impute such effects to a Being of superior rank, who had power over him as a Potter has over his clay.

There is no reason therefore to think that, from any notices within himself, such a Being would be able to form any conception of an external world; or, if he could, that an inquiry into the cause of his Sensations and Ideas would lead him into the conception of such a world and the belief of its existence.

It appears therefore that in order to refute Berkeley's scheme, | and consequently to lay a foundation for the system of Materialism, something more solid is wanted than the probability offered by Dr. Priestley.

III. I apprehend that to support the system of Materialism, the capital doctrines of Locke's system must be given up, and refuted.

To these doctrines Dr. Priestley has hitherto shewn a strong attachment; but if they leave no foundation for Materialism, the defenders of it must refute them.

In the preface to the third volume of his Institutes, page 21, He says, "I have no doubt, however, but the time will certainly come, when the general principles of Hartley, as well as of Locke, will be fully established, and when every contrary hypothesis will be forgotten."

In the Examination of Reid, Beattie and Oswald, after giving the outline of Locke's system, He adds, *pages 5 and 6*, "I think he has been hasty in concluding that there is some other source of our Ideas besides the external senses; but the rest of his system appears to me, and others, to be the cornerstone of all just and rational knowledge of ourselves. This solid foundation, however, has been attempted to be overturned by a set of pretended Philosophers, of whom the most conspicuous and assuming is Dr. Reid, Professor of Moral Philosophy in the University of Glasgow." And in a passage already quoted he acknowledges, 'That the hypothesis of there being no external world, is by no means so shocking to his understanding as the supposition that we really perceive things that are external to us."

The reasons why I conceive Mr. Locke's system concerning Ideas to be subversive of Materialism are two. First, because it leaves us no idea of substance, and Secondly, because Berkeley's system is the necessary consequence of it.

Mr. Locke saw clearly, and has expressly said, that we neither have nor can have any idea of substance, either by sensation or by reflection. And indeed nothing is more evident. For, by our senses we perceive only the qualities of bodies, | but not their substance; by consciousness we perceive only the operations of our minds, but not their substance; and there is not a third avenue according to Locke's system by which this idea can enter. Substance therefore is a word without an Idea, that is, without a meaning. And to say that Soul and Body are one Substance, or that they are two Substances, must be propositions that have no meaning.

I know there are many passages in Locke's Essay that imply our having an idea of Substance. When he came to consider our ideas in detail, it appeared that Substances occupy too large a share to be overlooked. Accordingly he divides them into three classes, to wit, the Ideas of Substances, of Modes, and of Relations, and treats of each particularly. One would think that, according to his principles, there can be no idea of Substances.

Whatever way we take to reconcile these things it is certain that no idea of Substance can be admitted, without overturning a fundamental article of Locke's system. The only account I can make of this apparent inconsistence is, that in candid minds, such as Mr. Locke's certainly was, selfevident truths, when, by some favourite hypothesis, they are shut out at one door, are apt to make good their entrance by another.

That Berkeley's system is the necessary consequence of Mr. Locke's appears to me very evident.

We are taught by Des Cartes, by Locke, and by all their followers, that all the evidence we can have of things external, must be drawn by reasoning from the Sensations and Ideas we are conscious of in ourselves, or, as it is very properly expressed by Dr. Priestley, that we really perceive nothing external, but judge of all things that are without ourselves by notices perceived within.

This principle, which I take to be the cornerstone of Locke's system, Berkeley assumes as an axiom; and draws his conclusion from it, by reasoning so clear and cogent, that they must stand or fall together. All attempts to disjoin them have been, and for ever will be vain. And therefore to establish the existence of Matter, and thereby lay a foundation | for Materialism, this principle of Locke's System must be refuted.

Chap. II. Of Newton's Rules of Philosophizing.

Dr. Priestley is sorry to have occasion to begin his Disquisitions on the nature of *Matter and Spirit*, with desiring his Reader to recur to the universally received *Rules of Philosophizing*, such as are laid down by Sir Isaac Newton at the beginning of his third book of *Principia*.

"Though, says he, we have followed these Rules pretty closely in other philosophical researches, it appears to me that we have, without any reason in the world, entirely deserted them in this. We have suffered ourselves to be guided by them in our inquiries into the causes of *particular appearances* in Nature, but have formed our notions, with respect to the most *general* and *comprehensive principles* of human knowledge, without the least regard, nay in direct contradiction to them. For my own part, he adds, I profess an uniform and rigorous adherence to them; but then I must require that my own reasoning be tried by this and by no other test.

"The first of these Rules as laid down by Sir Isaac Newton is, That we are to *admit no more causes of things than are sufficient to explain appearances*: And the second is, That *to the same effects we must, as far as possible, assign the same causes*."

To adhere uniformly and rigorously to Sir Isaac Newton's Rules, it is necessary to understand them perfectly, and to interpret them more justly than is done in the passage we have quoted. The plainest rules may be misapplied, and misinterpreted, if due attention be not given to their design, and to the words by which they are expressed.

Before we consider the interpretation which is here given of these rules, it may be proper, for the sake of those who are less acquainted with the *Principia* of Newton, to give some account of that work, of its phraseology, and of the intention and application of his Rules of Philosophizing.

That grand work is a book of *Natural Philosophy*, as that word is now used, that is, of the Philosophy of Matter, | or of the material system. And the whole business of Natural Philosophy, as that great Author teaches, is comprised in these two things, first to discover the laws of Nature, and secondly to apply the laws of Nature to explain the Phaenomena of Nature.

A Phaenomenon of Nature is said to be *explained* or *accounted for*, when it is shewn to be the necessary consequence of a known law of Nature. In that branch of Natural Philosophy which explains phaenomena, there is no need of any other Rules of Philosophizing, than the rules of demonstrative reasoning. For a phaenomenon which is not demonstrated to be a necessary consequence of a law of Nature, is not explained or accounted for.

In Natural Philosophy, the law of Nature, of which any Phaenomenon is the consequence, is called the *Cause* of that Phaenomenon. And when Sir Isaac Newton speaks of the *Causes* of *phaenomena*; the *Causes* of *natural effects*, or of *natural things*, he means, not the efficient cause, but, the law of Nature according to which the efficient cause operates in the production of those phaenomena.

Phaenomena of Nature, *Natural effects*, or *Natural things*, are so called, to distinguish them from the effects of human will and power. The voluntary actions of men are never called *phaenomena of Nature*, or *natural effects*. They may be called *civil* or *moral* phaenomena, being regulated by civil or moral laws which are often transgressed. But the laws of Nature, of which Natural Philosophy treats, being not only enacted by the Author of Nature, but put in execution, either by his immediate operation, or by means appointed by him, are never transgressed, unless he should miraculously interpose to suspend or change them.

In the two first books of the *Principia*, to which the title of that work most properly corresponds, the Author lays down, as Axioms, those laws of Nature regarding the motion of bodies, which had before been discovered by Philosophers, and sufficiently confirmed by experiments: And from those Axioms he demonstrates, by Mathematical Reasoning, the motions of bodies that must take place, on the supposition of their being acted upon by various centripetal and centrifugal forces; and, on the contrary, the forces that may be inferred from various supposed motions, either in an unresisting, or in a resisting medium.

These two books therefore, are very properly called *The Mathematical Principles of Natural Philosophy*. Their intention is not to shew what really exists; but to demonstrate the necessary consequences of certain suppositions.

The third Book, which is intitled *De Systemate Mundi*, is not hypothetical. Its intention is to shew what is the real constitution of our planetary system.

For this purpose he first lays down, and by a proper induction from facts, proves the truth of a law of Nature formerly unknown (to wit the general law of Gravitation) and then applies this, and the known laws of Nature, to explain or account for the phaenomena of our planetary system.

The discovery of this law of Nature, and the train of consequences demonstratively deduced from it, was the grand effort of that wonderful Genius, and is, undoubtedly, the greatest discovery ever made in Natural Philosophy.

The laws of Nature are not capable of demonstrative proof; but must be drawn from the phaenomena by just induction, like to that by which we deduce the grammatical rules of a Language from the Language itself; and it requires great caution, to distinguish that induction which is the proper evidence of a law of Nature, from plausible conjectures and hypotheses, by which the Philosophy of Nature has in all ages been contaminated.

Lord Bacon, at great length, and with wonderful sagacity laid down the rules of just induction, before the world had ever seen a proper example of it.

Sir Isaac Newton treading in the steps of this great Reformer in Philosophy, has given two of the first and the noblest exam|ples of this chaste induction, the first in the third book of the *Principia*, and the second in his *Opticks*. He has likewise comprehended the substance of Lord Bacon's rules in three selfevident Axioms, which he calls the Rules of Philosophizing. With these he begins the third book of the *Principia*; and they are evidently intended as a test, by which the truth of that law of Nature, which he discovered, may be tried, and which may equally well serve for the trial of all other real or pretended Laws of Nature, to confirm those that are true, and to reprobate those that are fictitious.

He had learned from Lord Bacon that sound natural Philosophy is nothing but an interpretation of the great volume of Nature, without adding any thing from our fancy to what it speaks, and his rules of Philosophizing flow naturally from this sentiment.

In every case therefore, when we would examine a law of Nature, or cause of natural phaenomena, Sir Isaac Newton's rules of Philosophizing are the proper test, and no cause of natural phaenomena ought to be admitted which they exclude, nor any rejected, which they authorise.

The constitution of the human mind, and all that necessarily flows from its constitution, though it does not belong to what is now called *Natural* Philosophy, may justly be considered as part of the great volume of Nature. Being, therefore, the work of Nature, its powers and faculties, their extent and limits, their growth and decline, and their connection with the state of the body, may, not improperly, be called phaenomena of Nature. And as far as these phaenomena can, by just induction, be reduced to general laws, such laws may properly be called laws of Nature.

Whether Sir Isaac Newton, in his rules of Philosophizing, had in his view the natural phaenomena of the mind, or not, does not appear; but, it is evident, that the reason of them extends to these, as well as to the phaenomena of the material system; and therefore they may be applied to both with equal | propriety, and ought to be adhered to with equal strictness.

But it is to be observed, that the voluntary actions of men can in no case be called natural phaenomena, or be considered as regulated by the physical laws of Nature. Our voluntary actions are subjected to moral, but not to physical laws. The moral as well as the physical laws of Nature are enacted by the great Author of Nature, but they are essentially different. The physical laws of nature are the rules by which the Deity himself acts in his government of the world, and, therefore, they are never transgressed. Moral laws are the laws which as the supreme Lawgiver he prescribes to his reasonable creatures for their conduct, which, indeed, ought always to be obeyed, but, in fact, are often transgressed. To suppose that his physical laws impose a necessity of transgressing his moral laws, would be to suppose that the wise and righteous Governor of the world has subjected us to contradictory laws, which is absurd.

There are many important branches of human knowledge, to which Sir Isaac Newton's rules of Philosophizing have no relation, and to which they can with no propriety be applied. Such are

Morals, Jurisprudence, Natural Theology, and the abstract Sciences of Mathematicks and Metaphysicks; because in none of those Sciences do we investigate the physical laws of Nature. There is therefore no reason to regret that these branches of knowledge have been pursued without regard to them: But it were much to be wished that, in inquiries into particular appearances in Nature, they had been as strictly followed as Dr. Priestley conceives.

Having endeavoured to explain the general intention of Sir Isaac Newton's rules of Philosophizing, what he means by *Phaenomena of Nature*, *natural things* and *natural effects*, what by the *Causes* of such effects, and what by *explaining* or *accounting for* them; having shewn to what branches of Philosophy his rules can, and to what they cannot be applied; let us compare the rules themselves with the interpretation which Dr. Priestley gives of them.

[17] "The first of these rules as laid down by Sir Isaac Newton, | he says, is that we are to *admit no more causes of things than are sufficient to explain appearances*; and the second is, that *to the same effects, we must, as far as possible, assign the same causes*."

These two rules as they are expressed by Newton admit of a literal interpretation, which is always best when the language will bear it. Rules of such a Writer as Sir Isaac Newton may be injured, but can never enlightened by a loose paraphrase.

The first rule as really laid down by Sir Isaac Newton is this, Of natural things no more causes ought to be admitted, than such as are both true and sufficient to explain their phaenomena. *Causas rerum naturalium non plures admitti debere, quam quae et verae sint et earum phaenomenis explicandis sufficiant.*

We may first observe that Newton's rule is restricted to natural things. Dr. Priestley extends it to things in general, without any restriction, as if the word *natural* had no meaning at all. This might be an oversight of a very inattentive reader, but it alters the meaning of the rule by extending it to things that were never meant by it.

But how shall we account for this, that when the rule requires two conditions of causes that are to be admitted; the first is suppressed altogether, and only the second mentioned. Did the Interpreter imagine that these two phrases, *such as are true and such as are sufficient to explain their appearances* meant one and the same thing, and that Sir Isaac in expressing a rule had been guilty of an useless tautology which needed his correction? If the words themselves, and the character of the Author had admitted this construction, it is precluded by the disjunctive particle *both*, which must imply two things and can never be applied to one.

If by *causes* in this rule, efficient causes were to be understood, there would be an impropriety in applying to such causes the attribute of truth, which can properly belong only to a proposition, or to what is expressed by a proposition. We shall hardly find such improprieties in the language of Sir Isaac Newton. When we consider that by *causes* he means, not efficient causes, but laws of Nature, real or pretended, the attribute of *truth* is applied to such causes with the greatest propriety; because laws of Nature, being general propositions, must be true or false. Sir Isaac Newton's rule requires that none be admitted but such as are true. But his Interpreter seems to have thought this condition unnecessary.

All the methods of Philosophizing, about the works of Nature, that ever have been used, may be reduced to two.

In the first, after a general survey of a certain class of phaenomena, the Philosopher sits down to conjecture from what causes, and how, they may be produced. Here, invention, the favourite power of the human mind, is set to work. A man who is blessed with this talent, hits upon an ingenious conjecture, which explains many of the phaenomena, perhaps all he has observed. Immediately, he concludes that, by his sagacity, he has discovered the Secret of Nature. He forms a system upon his hypothesis, which is fondly embraced and believed by his admirers. Time, however, brings to view phaenomena which contradict this hypothesis; it begins to lose its credit, and must give way to some new hypothesis which suits better with the phaenomena.

Thus one system triumphs for a time upon the ruins of those that came before it, and after a longer or a shorter reign, must undergo the fate of its predecessors; all being grounded upon the same false notion, that human wit and invention is sufficient to discover the art of Nature.

This method of Philosophizing, being best suited to the pride of human Genius, seems to have been followed, from the first origin of Philosophy, to the time of Lord Bacon. Whether the causes conjectured were intelligible, as in the system of Des Cartes, or unintelligible, as in that of Aristotle, the method of Philosophizing was the same, and is reducible to this Rule, as laid down by Dr. Priestley, that no more is required in causes to be admitted in Philosophy, but that they be sufficient to explain appearances.

Lord Bacon shewed that this method of Philosophizing is vain; that nothing solid in the Philosophy of Nature can ever be expected from conjecture and hypothesis, and that nothing is to be admitted, but what is deduced by juste reasoning from what our | senses testify. He pointed out the method by which laws of Nature, or Axioms (as he calls them) may, by just inductive reasoning, be drawn from observations and experiments.

Sir Isaac Newton adopted his sentiments, and never deviated from the tract he pointed out. His Rules of Philosophizing are formed upon Lord Bacon's principles; and he has given one example of their application in his *Principia*; and another in his Opticks; which furnish the best comment upon them.

In the third book of the *Principia*, before the law of Gravitation is applied to explain appearances, it is first proved by induction from the observations of Astronomers.

And in the *Opticks*, the laws of the Refraction and Reflection of light which he discovered, are first proved by his own experiments, before he applies them to explain appearances.

It is evident therefore, both from the words of this first rule, and from the method in which Sir Isaac Newton philosophized in consequence of it, that he did not conceive the explaining appearances to be the only condition required in a cause of natural things that ought to be admitted, and that he considered the proof of its truth as an essential condition. And when Dr. Priestley, professing to give us the Rules of Philosophizing *as laid down by Sir Isaac Newton*, suppresses or omits this condition; I know not how he can be acquitted from giving the name and authority of that great Philosopher, to a Rule, which is contrary to his express words, and contrary to the spirit of his Philosophy.

To represent this capital Rule, as if it gave a sanction to hypotheses, which have no evidence but that of explaining appearances, is to contradict its main design.

Sir Isaac Newton's second Rule of Philosophizing (says Dr. Priestley) is, *That to the same effects we must, as far as possible, assign the same causes*.

The Rule as laid down by Sir Isaac Newton is, Of natural effects

Part Three: Materialism

of the same kind, the causes are the same. *Effectuum naturalium eju⟨s⟩dem generis, eaedem sunt causae.*

Whether the Rule be mended, by putting, *the same effects*, for *effects of the same kind*, that is, an ambiguous expression for one that has no ambiguity, is not worth while to consider.

But first we must observe, that Newton restricts this Rule as well as the first to natural effects. His Interpreter extends both to effects in general, without restriction; which, as was before observed, gives a meaning to the Rule, that was never intended. |

[20] It may also be observed, that his taking away half of the first Rule, is compensated by making an addition to the second. For this clause, *as far as possible*, is purely an addition of the translator. If it have any meaning, it must imply, that the danger to be guarded against in applying this rule, is that of assigning different causes to effects which are acknowledged to be of the same kind. There is no danger of this to those who believe that there is one great Author of Nature, who governs the whole by uniform laws. The danger lies wholly on the contrary side, and is that of assigning natural effects to the same cause, on account of some apparent similitude, when more accurate attention would discover them to be of a different kind.

Men are naturally more prone to observe the similitude of effects, which may lead to the belief of their being of the same kind, than their differences, which might shew them to be of a different kind. The proper caution therefore with regard to this Rule is, not That we assign Effects to the same Cause *as far as is possible*, but that we be sure the effects be of the same kind before we assign them to the same cause.

This caution, though not expressed, seems to be insinuated by Sir Isaac Newton, by the examples which, for illustration of the Rule, he gives of effects of the same kind. *Such as*, says he, *Respiration in Men and in Brutes, the descent of stones in Europe and in America, light in the Sun and in a culinary fire, the reflection of light in the Earth and in the Planets.*

Sir Isaac Newton has given a third Rule of Philosophizing of which Dr. Priestley, for what reason I know not, makes no mention. It is, that "Qualities of Bodies which admit neither of increase nor diminution, and which are found to belong to all bodies on which we can make experiments, ought to be held as qualities of all bodies whatsoever".

By this Rule we are taught, what qualities, experience may lead us to conclude to be universal, or common to all bodies.

In that kind of induction which is the only proof we can have of a law of Nature, we draw a general conclusion from particular premises. This is contrary to the Rules of Syllogism, and, indeed, is not demonstrative proof; but it is the only proof we can have; and it is such a proof as we rest upon with perfect security in the common affairs of life. |

[21] That water will drown, and fire burn, are general propositions, which we learn by induction from particular facts or instances. Yet every man in his senses, is as much convinced of their truth as of any mathematical proposition. The induction, by which we are convinced of the truths now mentioned, and of many others of a like nature, is not carried on by Art or by Rules, but rather by habit; and by that principle of our constitution, by which we are led to expect that future events will be similar to those we have observed in time past in like circumstances.

When this expectation has been fulfilled in innumerable instances, it grows up into a perfect assurance.

One great cause of error, however, both in common life and in the philosophy of Nature, is, that men are prone to draw the general conclusion from too few instances, and to overlook, in instances apparently similar, those circumstances which shew them to be of a different kind.

From this source have arisen the notions of Omens, of Lucky and Unlucky Days, and many other vulgar Superstitions, to which Mankind have been prone in all ages.

From this source almost all the false theories in Natural Philosophy have arisen. They have been grounded upon induction, but upon a lame and imperfect induction, which is much more apt to lead into error than into truth.

Thus Des Cartes observing, as he tells us himself, how chaff is carried round in a tub of water in motion, and passively yields to all the motions of the fluid, concluded that the revolution of the heavenly bodies in their orbits is a phaenomenon of the same kind, and therefore that they are carried round in a fluid in which they float. And then, with a very fertile invention, he contrived a fluid fit for the purpose.

There is therefore an induction from particular instances to a general conclusion, which yields full conviction to men of understanding; but there is likewise an induction which is lame, imperfect, and extremely apt to lead into error. |

[22] To distinguish these two kinds of induction properly, and to fix the limit where one ends and the other begins, was a great *Desideratum*. It was a species of Logick, or of the art of Reasoning, which, as far as I know, had never been cultivated by any Philosopher, until it was undertaken by the great Lord Bacon, who, with a penetration truly wonderful, has taught us to distinguish compleat induction from that which is lame and imperfect.

Sir Isaac Newton, who, in his method of Philos(o)phizing, followed Bacon so closely as even sometimes to adopt his peculiar phraseology, has, in this third Rule of Philosophizing, expressed as much of the art of just induction as was necessary for confirming that law of Nature which he discovered. *Qualitates corporum quae intendi et remitti nequeunt, quaeque corporibus omnibus competunt in quibus experimenta instituere licet, pro qualitatibus corporum universorum habendae sunt.*

It is upon the principle of this Rule, as Newton observes, that we believe that all bodies are moveable, that all bodies are divisible, and that all bodies, on the earth's surface, gravitate towards the earth. And he shews that the same principle, compared with the observations of Astronomers, leads ⟨us⟩ to conclude that the law of Gravitation is common to all bodies whatsoever.

As this Rule particularly points out, upon what evidence we are to hold a quality of bodies to be universal, or to belong to all Matter, it is the more strange that it is totally overlooked by Dr. Priestley. He holds Matter to be universally endowed with inherent powers of Attraction and Repulsion, he holds that there is never any contact between the parts of Matter, and, that *Inertia*, which most Philosophers conceive to be common to all Matter, is no quality of Matter at all. In these opinions about qualities common to all Matter he differs from Sir Isaac Newton and from most Philosophers. He appeals to Sir Isaac Newton's rules as the test, by which alone these opinions are to be tried. It is therefore the more |

[23] strange that he should overlook the only rule of the three which teaches how we may, from induction, conclude a quality to be common to all bodies whatsoever.

From what has been said in this chapter I think it appears, that the rules of Philosophizing which Dr. Priestley has laid down as the rules of Sir Isaac Newton, are not the rules of Sir Isaac Newton; and that if he should adhere to the rules he has given as Sir Isaac Newton's, as uniformly and rigorously as he professes, he may be led wrong by them.

In the progress of the Disquisitions we are often told, that such a point must be admitted if we adhere to the established rules of Philosophizing, while it is left to the Reader to find how the established rules are to be applied to the point in hand. In other instances, the first or second of the rules is referred to, according to the lame interpretation which he had given of them.

Chap. III. Of the Solidity or Impenetrability of Matter.

In the introduction to the Disquisitions, we are told, that "It is maintained in that treatise that Matter is not that *inert* substance that it has been supposed to be; that powers of *attraction* or *repulsion* are necessary to its very being, and that no part of it appears to be *impenetrable* to other parts". He "therefore defines it to be, a substance possessed of the property of *extension*, and of powers of *attraction or repulsion*. And since it has never yet been asserted, that the powers of *sensation* and *thought* are incompatible with these, (*solidity*, or *impenetrability*, and consequently a *vis inertiae*, only, having been thought to be repugnant to them) he therefore maintains, that we have no reason to suppose, that there are in man two substances so distinct from each other as have been represented."

We see then, that the general properties the Author excludes as not belonging to Matter, are *Solidity* or *Impenetrability*, and a *Vis Inertiae*, and that, to supply the want of these, he ascribes to it powers of attraction or repulsion, as necessary to its very being. We may, first, consider the grounds on which he excludes Solidity or Impenetrability from being a property of Matter.

Solidity and Impenetrability, Dr. Priestley, I think justly, con|siders as the same property under different names. But it were to be wished that he had given a definition of this property, because *Solidity* has different meanings which it concerns his reasoning to distinguish.

In a popular sense, *Solidity* is opposed to *Fluidity*, and signifies a certain degree of hardness, or cohesion of the parts of a body.

In another popular sense, it is opposed to hollowness, or vacuity in the internal parts of a body. In neither of these popular meanings is *Solidity* a property of all bodies, far less an essential property.

There is also a mathematical sense of *Solidity*, in which it signifies extension in three dimensions. This belongs to space as well as to body.

There remains a philosophical meaning of *Solidity*, in which it is commonly believed to be an essential property of all bodies, and in this sense it means, that a body occupies its place in such a manner as necessarily to exclude all other bodies from occupying the same place at the same time.

This, and this only, is the *Solidity* or *Impenetrability*, about which Dr. Priestley and the Philosophers differ; they conceiving it to be an essential property of all Matter, and he conceiving that it is no property of Matter at all.

His arguments, therefore, must be restricted to this meaning of *Solidity*, and if in any instance their force reach only to *Solidity* taken in another sense, they are foreign to his purpose.

Dr. Priestley acknowledges, "that there are common appearances enow which must necessarily lead the Vulgar to the judgment he opposes. – But they are false appearances, and therefore have led to superficial and false judgments; judgments, which the real appearances will not authorize."

I acknowledge on the other hand, that I know no appearances to the senses, common or uncommon, which prove *Solidity* to be a necessary property of Matter. For, first, our Senses are too imperfect to determine the fact, that Matter is never penetrated by Matter. |

When gold is amalgamated with Mercury, the human eye cannot determine whether the Mercury penetrates the substance of the gold, or only is received into the interstices of its minute parts. So it is in all chymical combinations.

And secondly, if our senses did testify that, in fact, Matter is never penetrated by Matter, they cannot possibly testify that it is necessarily impenetrable. Matters of fact may be testified by the senses, but necessary truths are beyond their sphere.

It is probable, therefore, that the Philosophers who hold *Solidity* to be a necessary property of Matter, have been led to that opinion, not by the common appearances which the Author mentions, but by consequences which follow from the supposition of Matter being unsolid and penetrable, which appear to them absurd.

I shall mention some of these consequences, since those who deny Matter to be solid, ought either to adopt them, or shew that they do not follow from the opinion which they defend.

Suppose then, that two particles of Matter may occupy the same place at the same time. Neither of them having Solidity, they make one unsolid particle; a third, therefore, may be received into the same individual place; and, for the same reason, ten thousand more, or even the whole Matter of the Universe. There is, therefore, no extension in three dimensions so small, as that it cannot contain the whole matter of the Universe, and yet have the same capacity to contain more worlds, as when it was empty space. In a word, the capacity of the minutest atom of extension to hold matter within itself, is, really and without a figure, infinite. And this atom, once duly impregnated with Matter, may, without a miracle, without any new creation, bring forth from its fruitful womb ten thousand worlds.

Nay, this Atom, being still an extended substance, that is, having parts without parts, one part may penetrate another part, until it be reduced to a mathematical point, and have no extension. So that if Matter have not Solidity, extension is not a necessary property of it, since the whole Matter of the Universe may be contained in a mathematical point. |

Suppose again, three particles of Matter, equal in every respect, to occupy the same place at the same time. There were three distinct substances before this union. Are there now three, or only one? There is only one substance surely; for it has all the unity that we can conceive in a body. We cannot, even in imagination, ascribe number, where there is no distinction either of nature, or of time, or of place, or of any other circumstance. Thus three substances may become one without any annihilation, or one substance may become three without creation. Is there not here an Unity in Trinity, and Trinity in Unity, which, I think, Dr. Priestley holds to be an absurdity?

Once more, may we not, according to this system of Materialism, suppose two or three thinking and intelligent brains, perfectly similar, to exist in the same place and time. This is evidently possible upon the principles of this system. And would not this union be as mysterious, as that of Soul and Body, on the supposition of their being substances of a different kind; and as mysterious as other unions of Natures and Persons, which Dr. Priestley thinks to be absurd.

Upon the supposition of the Penetrability of Matter, I think, we might account for Transubstantiation, or, at least, for Consubstantiation; and, I hope, those who think themselves obliged to maintain these mysterious doctrines, will be thankful for a discovery, which affords them an aid that was much wanted.

Philosophers therefore have probably been led to think Solidity not only a property of Matter, but a necessary property, that they might avoid the absurd consequences of the contrary opinion, rather than from common appearances to the senses; For he must be a sorry Philosopher who does not understand, that though the senses may testify what *is*, they never can testify what is *necessary*.

Let us now consider the arguments brought to prove that Matter is not solid or impenetrable. I wish that I could state them more distinctly than I am able to do. But I think they are all drawn from these principles, That Matter is possessed of attracting and repelling powers, and of several spheres of them, one within another, That on account of the repelling power, the parts of Mat|ter can never, by any mechanical force, be brought into actual contact, but are kept at some distance both in the composition of bodies and in their collision, That there is nothing in or belonging to Matter, capable of resistence, besides its powers of repulsion, That taking away the power of attraction, Solidity itself vanishes, and not only so, but, without this power of attraction, the very substance of Matter vanishes, and is annihilated.

In these positions there are some things that ought to be granted, others, of which there appears no evidence, nor do I see that, upon the whole, there is any proof that Matter is not solid and impenetrable.

That there are various powers of Nature, by which bodies are made to accede to one another, or recede from one another, was clearly shewn by Sir Isaac Newton; some of these powers producing their sensible effects only upon the smaller parts of Matter, and at small distances, as in the cohesion of hard bodies, corpuscular attraction, the powers that operate in chymical processes, and those that act upon the rays of light in their emission, and in their reflexion, refraction, and inflexion; other powers exhibit sensible effects upon larger bodies, and at greater distances, as those of Magnetism, electricity, and gravitation.

It is likewise true, that in many instances of the collision of bodies, and of their pressure against one another, the parts which, from the imperfection of the human senses, were thought to touch, are found to be kept asunder by a repelling power.

These things being granted, let us consider what follows from them. Dr. Priestley thinks that we must, according to the rules of philosophizing, conclude, that there never is any real contact of the parts of Matter. This, I humbly think, is to put more in the conclusion than is in the premises; and I know no just rules of Philosophizing that authorize such a conclusion. It is certain that Sir Isaac Newton, who may be supposed to have understood his own rules, never thought that this conclusion could be drawn from the premises above mentioned.

[28] But Dr. Priestley thinks it demonstrable, "that the particles | of the hardest bodies do not touch, because they are brought nearer by cold, and removed farther from each other by heat."

This demonstration reaches only to the particles perceivable by our senses; that they do not touch it proves; but these particles must be composed of other particles, and whether they touch or not this demonstration leaves us quite in the dark.

There is a dilemma in Dr. Priestley's hypothesis, which I know not how to resolve, without admitting contact, or annihilating Matter altogether.

Every the least particle of Matter has various spheres of attraction and repulsion, one within another. Must there not therefore be an inmost sphere, which must either be a sphere of attraction or a sphere of repulsion?

Let us first suppose it a sphere of attraction. Does it not follow, that all the particles within this sphere will run into mutual contact? This, indeed, is what Sir Isaac Newton, and, I think, most Philosophers suppose, but we see it necessarily infers contact.

And if we suppose, with Dr. Priestley, that Matter is not solid nor impenetrable, it follows, that the particles within this sphere

Part Three: Materialism

of attraction will not only touch, but penetrate each other, until they be all attracted into a mathematical point.

To avoid this conclusion, and exclude all contact, let us take the supposition that the inmost sphere is a sphere of repulsion.

Upon this supposition, the matter contained within this sphere of repulsion will be a perfectly elastick fluid, its minutest parts being all separated from each other to a certain distance; and these parts must be unextended and indivisible. For whatever is extended has parts, and these parts, by their repulsive power, will divide and separate 'till no part be left that is either extended or divisible. Thus Matter will be composed, not of physical points, as Dr. Priestley holds, but of Mathematical points, according to the more consistent system of Boscowich. And of such elastick fluids, retaining still their elastick force, every body, even the hardest, must be composed.

[29] It will be difficult to reconcile this second supposition with | Dr. Priestley's argument to prove, that without the power of attraction the very substance of Matter would be annihilated. Without a power of attraction, he says, "every particle would fall from each other, and be dispersed. And this being true of the ultimate particles, as well as of gross bodies, the consequence must be, that the whole substance will absolutely vanish."

If the want of a power of attraction have this effect, a power of repulsion cannot have a less effect. Or if the particles may be dispersed to the extent of the sphere of repulsion without losing their existence, it will be difficult to shew that they must be annihilated by being dispersed to a greater distance.

The solution of this dilemma I must leave to Dr. Priestley.

To return to contact in the parts of Matter, I think that what Dr. Priestley has said does not prove that there is no such contact. But supposing that there were no contact, will it follow that Matter is not solid or impenetrable? I think it will not.

If Matter never be touched by Matter, a certain consequence is, that it is never penetrated; for in order to be penetrated, it must first be touched. All therefore that can be inferred from its not being touched is, that it can never be put to the test whether it be penetrable or not.

As far as I can comprehend Dr. Priestley's reasoning, it amounts to this; That the sole reason why Philosophers have concluded matter to be solid and impenetrable, is grounded on the supposition of a real contact in the parts of matter, and that resistance is caused by that contact; but that supposition is false. Therefore &c

The conclusion from these premises, supposing both to be granted, is, that those Philosophers have reasoned from a false supposition; and therefore that their reasoning does not prove Matter to be solid and impenetrable; but neither does it prove the contrary. Bad reasoning on one side of the question cannot supply the want of good reasons on the other. All the effect it can have is to leave the question a moot point. Dr. Priestley therefore might say "that by such false reasoning, he is not authorised, as a Philosopher, to believe matter to be solid"; but when he adds, that "on the contrary he is obliged to deny that it has any such property", this is contrary to the rules of reasoning; for no man is obliged to deny a proposition, because a bad argument has been used to prove it to be true. | If a man were to regulate his Belief by this rule, he must believe nothing at all; for he will not find an article in his Creed for which a bad argument has not been advanced.

I have endeavoured to shew that Philosophers may, very probably, have had reasons for believing matter to be solid or impenetrable, which do not go upon the supposition of a real contact in matter, and of its resistance depending upon contact. Be this as it will, bad reasoning proves only the weakness of those who use it, but with regard to the matter in which it is employed, it neither proves nor disproves any thing.

Whether the resistance of matter to any change of its motion be owing solely to its powers of repulsion will be considered in the next chapter.

Two of Dr. Priestley's principles from which he argues against the solidity of matter remain, and are both taken from the attracting powers, to wit, that taking away the power of attraction solidity itself vanishes, and that without this power matter would be annihilated.

With regard to these two principles we might first observe, that, if they were true, they would furnish us with a proof that matter is necessarily and essentially solid. For if the power of attraction be essential to matter, which he affirms, and solidity be the consequence of the power of attraction, which he also affirms, it follows necessarily, that solidity is essential to matter. But though this argument be good *ad hominem*, it is in itself good for nothing; for neither the solidity of matter, nor its existence, have the least dependence upon powers of attraction.

We observed before, that there is a popular meaning of solidity, in which it is opposed to fluidity. No Philosopher ever maintained this solidity to be essential to matter. It belongs to some bodies but not to all, nor even to the same body at all times. That *Solidity* taken in this sense depends upon powers of attraction, is granted.

There is also a popular meaning of penetrability, corresponding to the popular meaning of solidity. We say, that fluid and soft bodies are easily penetrated; hard bodies, with great difficulty. But this penetration means, not that both bodies occupy the same place at | the same time, but only that the penetrating body enters into the pores of the body penetrated, or by displacing some of its parts, occupies the place which they occupied before. With regard to this solidity and penetrability, there is no difference between Dr. Priestley and the Philosophers.

The solidity which Philosophers maintain, and Dr. Priestley denies to belong to matter, is, that it occupies its place so as to exclude other matter from occupying the same place at the same time. And the question before us at present is, Whether this property of matter depends upon powers of attraction?

It is impossible it should. And Dr. Priestley would have immediately perceived this, if he had not imposed upon himself and his reader by the ambiguity of the word solidity; taking it as it means hardness or cohesion of parts.

That he did so is evident, because the reason he gives why a solid atom cannot retain its solidity without an attraction of its parts, is, that without such mutual attraction of the parts, "this atom will be entirely destitute of compactness and hardness, which is necessary to its being impenetrable, and, that every particle will fall from every other particle and be dispersed."

But, granting that it is entirely destitute of compactness and hardness, and that its parts fall asunder and are dispersed; it may still retain perfect solidity or impenetrability, by excluding other matter from occupying the same place at the same time.

This is the fallacious argument by which he would prove, "that a power of attraction is necessary to the solidity of matter, that whatever solidity any body has, is the consequence of its being possessed of certain powers," and that "taking away attraction solidity itself vanishes." The fallacy is owing to the ambiguous meaning of the word *solidity*.

The second principle he advances, with regard to attraction, is, that, without it, matter itself must cease to exist and be annihilated. For which, this reason is given, that without attraction no substance can retain any form, shape or figure.

To this it may be answered, that, without attraction, a body will retain its form, till, by some impressed force, its form be changed; and, with attraction, it will retain it no longer.

But that the change of its form, be it ever so frequent, should annihilate it, is a new doctrine, and very strange. It supposes that the particles of matter, by being much agitated, will be driven out of existence.

In elastick fluids, with which Dr. Priestley has been much conversant, instead of a mutual attraction of their particles there is a mutual repulsion, which, surely, makes them less apt to retain their form, than if they wanted both. If the want of power to retain their form produced annihilation, there could never be any such thing as an elastick fluid.

We have referred to the next chapter the consideration of that position of Dr. Priestley, That there is nothing in or belonging to matter capable of resistance, besides its powers of repulsion. All that is here to be observed is, that supposing this position to be true, it can afford no argument against the solidity of matter. For as we have shewn that powers of repulsion are perfectly consistent with the solidity of matter, every effect of those powers must also be consistent with it.

In this chapter we have attempted only to defend the solidity or impenetrability of matter against the arguments of Dr. Priestley, by shewing, That the powers of attraction and repulsion do not prove matter to be penetrable, That they do not prove that there is no contact of the parts of matter, and, That even if they did, it would not follow, that matter is not solid and impenetrable, That powers of attraction are not necessary to the solidity of matter, and, That if those powers were withdrawn, it does not follow, that matter would be annihilated.

Chap. IV. Of the Inertia of Matter.

The force which a body opposes to any change of its state either of motion or of rest, and the force with which a body tends forward when it is put in motion, are conceived by Philosophers to be the same force, under different considerations; and therefore they have given to both the name of *Inertia* or *vis inertiae*. The last is sometimes called *vis insita*, to distinguish it from impressed forces, which have some external cause.

The indifference of matter to motion or rest, and the *vis inertiae* of matter, are both laid down by Sir Isaac Newton, with his usual accuracy and distinctness, in the three laws of motion, which he prefixed to the first book of the *Principia*; and which in the | style of Lord Bacon, he calls *Axioms*; meaning by that name, not that they are selfevident truths, but that they are laws of Nature, which, by preceeding Philosophers, had been sufficiently established by inductive proof; so that they might be assumed as known and certain, and applied either to the explication of phaenomena, or to the investigation of other laws of Nature.

The first is, That every body perseveres in its state either of motion or of rest, unless so far as, by impressed forces, it is made to change it.

The second, That the change of motion is proportional to the motive force impressed, and is made in the right line in which that force is impressed.

The third, That there is a reaction equal and contrary to every action of one body upon another; or, that the actions of two bodies on one another are always equal and in contrary directions.

Bodies are said to act one upon another, by pressing, by striking, by attracting, or by repelling; and in all these cases, the action is mutual, and equal; both are pressed, or struck, or attracted, or repelled with equal motive force, and in contrary directions.

These laws of Nature express all that is meant by the indifference of matter to motion or rest, and all that is meant by the *inertia*, or *vis inertiae* of matter; and they are so well established by the experience of ages, that it must appear very strange, that any one has given attention to Natural Philosophy should call them in question. All the principles of Mechanicks, and of every branch of Natural

Philosophy that relates to motion are built upon them, and must be overturned from the foundation if they be false.

In the infinite variety of Machinery, simple or complex, which human art has contrived, these laws have never been found to fail in a single instance. And it deserves to be observed that they take place, not only in every motion of inanimated bodies, but in every motion of animal bodies that falls within the reach of our senses.

Every part of the human body, and even the brain itself, which, according to this system of Materialism, is the thinking part, appears, in every experiment we can make upon it, to be possessed of the *vis inertiae* and subject to the laws of motion. |

Every motion of the human body perceivable by our senses, whether voluntary or involuntary, is caused by a force impressed by the contraction or relaxation of muscular fibres, and bears exact proportion to that force.

Every motion is accompanied with a reaction which is equal and in a contrary direction; and even the muscular fibres themselves are subject to this law.

But, by what force these fibres are contracted and relaxed, either in voluntary or involuntary motions, Philosophy has never discovered. We see that some influence of the nerves is necessary; but what kind of influence it is, or how it operates, we are perfectly ignorant.

The impressed force therefore, by which the muscular fibres are contracted or relaxed, is a phaenomenon not explained or accounted for by the laws of motion, and perhaps never will be. It is here that Mechanism ends, at least human wisdom has never been able to carry it farther; and some cause must operate that has the power of beginning motion.

To ascribe the power of beginning motion to the muscular fibres themselves, or to the nerves, or to the brain (for we have equal reason for ascribing this power to any one of the three or to them all) would be, first, to contradict the first law of motion, and secondly, to contradict Sir Isaac Newton's first rule of philosophizing, by introducing a law of Nature, which is neither proved to be true, nor explains the only phaenomenon for which it was invented. For, supposing the brain to have the power of beginning motion in itself, no man can shew how this power can contract or relax a muscular fibre, with which it has no communication but by the intervention of a slender nerve.

As, therefore, even a power in the brain to begin motion does not explain muscular motion, he that maintains the brain to be the efficient cause of that phaenomenon, must say, that this cause produces its effect by an occult quality of which we see no evidence in any other operation of Nature.

This indeed is a very ancient way of explaining appearances; but it is very different from Sir Isaac Newton's.

[35] To return to the Inertia of Matter. It is found by experience | in every part of matter on which we can make experiments, it is neither capable of increase nor decrease, being always precisely in proportion to the quantity of matter; and therefore, according to Sir Isaac Newton's third rule of philosophizing, (which Dr. Priestley has omitted to mention) ought to be held as a quality belonging to all bodies.

The rule is this, *Qualities of bodies which can neither be increased nor diminished, and which belong to all bodies on which we can make experiments, are to be held as qualities of all bodies whatsoever.* This rule of philosophizing applies to the inertia of matter as exactly as if it had been made for the purpose.

This property of matter Dr. Priestley thinks unfriendly to his system of Materialism, and therefore maintains, "that matter is not that inert substance that it has been supposed to be." But I do not find that he has offered any other argument against the *inertia* of matter but that "powers of attraction *or* repulsion are necessary to its very being."

I take him to mean, that one of these, to wit powers of attraction, are necessary to the being of matter, for I do not find any argument to prove that powers of repulsion are necessary to its being.

In the last chapter we considered his argument for proving that powers of attraction are necessary to the existence of matter; but as Sir Isaac Newton held powers of attraction and repulsion to be consistent with the inertia of matter, the inconsistence of inertia with those powers ought to have been shewn.

Perhaps the reason why Dr. Priestley, after reasoning so copiously against the solidity of matter through two sections, has said so little against its inertia, may be that he considered these two properties of matter as necessarily connected, so that they must stand or fall together; and that having, as he conceives, proved that matter is not solid, it follows of course that it is not inert. For in his introduction he says, that matter has been said to be possessed of solidity or impenetrability, and *consequently of a vis inertiae*, and again, that solidity or impenetrability and *consequently a vis inertiae* only, | have been thought incompatible with sensation and thought.

Upon this I shall only observe, that I apprehend it will be very difficult to prove that the inertia of matter is a necessary consequence of its solidity, and in order to compleat his argument it is necessary to prove that if it be not solid it cannot be inert.

Arguments drawn from experience to prove the solidity of matter, may perhaps be evaded by recurring to a repelling power; but I conceive it impossible to evade the force of the arguments drawn from experience to prove the inertia of matter.

Dr. Priestley indeed seems to think, that the phaenomena which we resolve into the inertia of matter will follow from a repelling power as well as those that are commonly imputed to the solidity of matter, and that all the three laws of motion may be accounted for by a repelling power.

Illustrations page 236, "It is said, that if there is not what has been termed a *vis inertiae* in matter, the foundation of the Newtonean Philosophy is overturned; for that the three laws of motion, laid down by Sir Isaac Newton, in the beginning of his Principia, have no meaning on any other supposition.

I answer, that these laws of motion are founded on certain facts, which result as easily from my hypothesis, as from the common one. It is an undoubted fact that every body perserveres in a state of rest or motion, till it be compelled to change that state by some external force, which is the first of the three laws, and the foundation of the other two. But this will follow just as well, upon the supposition of that mutual action between two bodies, taking place at any given distance from their surface."

In this passage Dr. Priestley, I think, acknowledges, that the three laws of motion are justly founded upon facts or phaenomena, and consequently that they are invariably observed in the course of Nature. But he thinks that these laws follow from a power of repulsion in bodies, without supposing them inert.

Upon this passage we may observe, First, that if the laws of motion be founded on certain facts, it follows that the inertia of matter is founded on certain facts. For the laws of motion express all that is meant by the inertia of matter as fully and as clearly as words can express any thing. And it is impossible, without | inconsistency, to admit the laws of motion and to deny the inertia of matter.

Secondly, if the laws of motion follow from a power of repulsion the consequence of this position is, not that matter is not inert, but that a power of repulsion is the cause of the inertia of matter.

Sir Isaac Newton, though he perceived that there are powers of repulsion between the parts of matter, had not the sagacity to discern that the laws of motion, and consequently the inertia of matter, is the result of powers of repulsion; and therefore he pretends not to derive the laws of motion from any cause but lays them down as facts confirmed by universal experience. The discovery of their cause was, it seems, reserved to Dr. Priestley, and indeed if it be true, it is an important discovery in Natural Philosophy.

Whether a power of repulsion in matter that has no inertia will produce the three laws of motion, is a point that admits of mathematical reasoning, and therefore may be determined with certainty.

Suppose, first, a body already in motion. If it have *inertia* it must have a *vis insita* proportioned to its quantity of matter and velocity. By this *vis insita* it goes on uniformly in a streight line, without any diminution of its force or velocity. The repelling power can have no power to continue it in motion.

Thus we see that a repelling power without inertia cannot produce the first law of motion; and that if matter be inert, whether it have a repelling power or not, that law will take place.

Let us now consider the change of motion in the collision of bodies, which is regulated according to the second law of motion.

Suppose the body A, with a certain velocity, to move directly against the body B at rest. If both are inert, the change of motion will be precisely what the laws of motion require, whether there be a repelling power or not. A repelling power may hinder actual contact, and make the change of motion take place a moment sooner than it would have done if there had been no repelling power; but these effects cannot be perceived by our senses. The change of motion and every effect of the collision perceivable by our senses will be the same whether there be a power | of repulsion or not. If now the repelling power continue, but the inertia of the body B be withdrawn: Will the effect of the collision be the same as before? This is impossible. The body B together with its repelling power, having no inertia will fly upon the approach of the body A, like its shadow, without making the least resistence, and consequently without diminishing in the least the velocity of A. And the effect will be the same, though the body A be ever so small, and B ever so great. For, without inertia, the whole matter of the universe will require as little expence of force to move it, as an atom; and therefore the same impressed force may produce either a small or a great change of motion, which contradicts the second law of motion.

A little attention might have satisfied Dr. Priestley that a power of repulsion in a body without inertia can make no resistence to any change of motion. The least force makes the power of repulsion and its body along with it to flee as fast as it is pursued without the least resistance. It has no power in itself to keep its place, and there is nothing behind it to support it, and therefore it can make no resistance.

It is demonstrable therefore that the three laws of motion will not follow from the supposition of a repelling force, in matter that is not inert. Nay so little dependance have those laws upon a repelling power, that whether there be such a power or not, the laws of motion will be the very same as far as the human senses can perceive. Accordingly we find that, before a repelling force in matter was known or suspected, when it was believed that contact took place on every collision of bodies, the laws of the collision of bodies were by demonstrative reasoning deduced from the inertia of matter, and found perfectly agreeable to experience; and now that repelling forces are admitted, the laws of the collision of bodies whether hard, soft or elastick, remain without any change; and are explained by the same principle without taking into consideration a repelling force.

Upon the whole the three laws of motion have all the evidence | that experience can give to a general proposition, and the inertia of matter must have the same evidence, because it only comprehends in one word what the three laws of motion express in detail.

If therefore Sir Isaac Newton's rules of philosophizing have authority; if, to use the words of Dr. Priestley concerning them, "so long as we follow these maxims, we may be confident that we walk on sure ground; but the moment we depart from them, we wander in the regions of mere fancy"; if this be so, it follows that all bodies are inert, and consequently have no inherent power of beginning motion.

That a Materialist should be unwilling to admit the inertia of matter, is not strange; but that, denying the *inertia* of matter, and affirming that it has inherent attractive and repulsive powers, he should pretend to build his system upon Sir Isaac Newton's rules of philosophizing, and require that it be tried by that test only; of this I can make no other account, but that of the three rules of philosophizing he misunderstood the two first and overlooked the last.

Chap. V. Of Powers of Attraction and Repulsion in Matter.

We have granted to Dr. Priestley that, in the collision of bodies, actual contact may be prevented by a power of repulsion, and likewise that, without powers of attraction, bodies could have no hardness or coherence of their parts. Every body would have that perfect fluidity which writers suppose in the theory of Hydrostaticks.

An important question remains, to wit, Whether these powers are inherent in matter as their subject, or whether they be not impressed forces, by which matter does not act, but is acted upon; does not attract and repel, but is attracted and repelled.

It is one thing to say, that the parts of matter are kept from contact by a power of repulsion, and that they cohere by a power of attraction; it is another thing to say, that matter in a strict sense is possessed of those powers. The first is very agreeable to Sir Isaac Newton's philosophy; | the last directly contradicts it. The powers of attraction and repulsion are always by that Philosopher called *impressed forces*, and by that name are distinguished from the *vis insita*, which he conceives to be inherent in matter.

Every power must be in some subject, in which it inheres, and which in the strictest sense is possessed of it. If matter itself be the subject of the powers of attraction and repulsion, it may in the strictest sense be said to be possessed of those powers; and this undoubtedly is what Dr. Priestley means by that phrase. For first, he holds those powers to be essential to matter, and surely matter cannot be possessed of them in any sense more strict than when they make a part of its essence; and secondly, he conceives these powers will "remove from matter the *odium* which has hitherto lain upon it from its supposed necessary property of solidity, inertness, and sluggishness; as from this circumstance only the baseness and imperfection which have been ascribed to it are derived. Since matter has in fact no properties but those of attraction and repulsion, it ought to rise in our esteem, as making a nearer approach to the nature of spiritual and immaterial beings, as we have been taught to call those which are opposed to gross matter." Disquisitions pag 17.

Now if those powers be not in matter but in some other subject, they prove no more than what all men always believed, to wit, that matter is capable of being moved by an impressed force; and whatever *odium* lay upon matter from its supposed inertness and sluggishness must remain, nor does it make any nearer approach to the nature of spiritual and immaterial beings.

We are therefore to consider whether it be proved that those powers of nature by which the parts of matter accede to each other or recede from each other, are really inherent in matter, so that it is not as a passive subject attracted and repelled, but in the strictest sense attracts and repels other matter.

[41] That we may not be perplexed by the multitude of attracting and repelling powers, let us confine ourselves to that | of gravitation, since what can be said of the cause of gravitation may easily be applied to all powers either of attraction or of repulsion.

When two bodies tend towards each other by gravitation, we perceive the effect, and we know that it must have a cause; but the cause and the manner of its operation is invisible and unknown. Whether the cause acts by attraction, or by propulsion, or in some way of which he have no conception, we know not. Whether the cause be matter, or mind, or something different from both, of which we have no conception, we know not.

When we examine the effect, we may discover some things relating to it. For instance, that gravitation is common to all bodies on which we can make experiments; that, at the same distance from the body to which it tends, it is precisely in proportion to the quantity of matter in the gravitating body; and that at different distances its force increases in the same proportion as the square of the distance decreases. These are the noble discoveries of Sir Isaac Newton; justly deduced from phaenomena according to his rules of philosophizing. But these discoveries, however important in themselves, carry us not a step towards the discovery of the cause of gravitation, or of the manner in which that cause operates. This has hitherto been involved in impenetrable darkness.

In this state of darkness, there are two ways that have been taken.

Some ing⟨e⟩nuously confess their ignorance, and say with Sir Isaac Newton, *Rationem harum gravitatis proprietatum ex phaenomenis non potui deducere; et hypotheses non fingo*. They may still to avoid circumlocution call it *attraction*, and refer it to some center towards which it is directed; but after giving repeated cautions that they mean only to express the effect, and are quite ignorant of the cause, their meaning, by the candid and attentive, cannot be mistaken.

There are other persons who take another course.

Rather than confess their ignorance, they chuse to grope in the dark till they hit upon some conjecture or hypothesis, which they conceive may be the cause of gravitation. After this they find little difficulty in persuading themselves that it is the true cause. |

Thus Dr. Priestley conjectures that all the effects of gravitation which we observe, are produced by an inherent quality of matter, by which every part of it attracts every other part, at all distances, at least to the utmost bounds of our solar system.

This, if true, is a new law of Nature, and therefore, according to Sir Isaac Newton's rules of philosophizing, its truth should first be proved, and then it may be applied to explain phaenomena.

Suppose the bodies A and B to gravitate towards each other; what is the cause of this gravitation? Dr. Priestley says, that A by an inherent power attracts B, and B by an inherent power attracts A, and these two powers are in all cases precisely equal, however unequal the bodies may be. Another man may say, it is not so; but the body A by an inherent power moves itself towards B; and B by an inherent power moves itself towards A with precisely the same force. A third person may think it a more simple hypothesis, that some invisible matter in the common center of gravity of the two bodies attracts both with equal force. And a fourth, that there is no attraction at all, but propulsion by matter external to both bodies.

These four hypotheses may all be false, and only one of them can be true. Suppose we take it for granted that some one of them is the true cause (which is contrary to Sir Isaac Newton's rules of philosophizing) how shall we determine the preference?

I do not find that Dr. Priestley has shewn that his own rules, far less those of Sir Isaac Newton, determine this point. No proof therefore has been given of this cause of natural things.

And as to the explication of phaenomena, this cause is of no use. For all the phaenomena referred to gravitation, have already been sufficiently explained by the properties of gravitation above mentioned, which Sir Isaac Newton discovered without pretending to any knowledge of its cause.

We are therefore required to admit a cause of natural things, which is neither proved to be the true cause, nor ex|plains any appearance, and that by one who *professes an uniform and rigorous adherence to Sir Isaac Newton's rules of philosophizing, and requires that his reasoning be tried by this, and by no other test.*

Besides the want of evidence in this hypothesis, and its inutility to explain any of the phaenomena of Nature, there are other objections that press heavily upon it.

It implies that every part of unthinking matter, by an inherent power, acts so wisely and regularly as to give an impulse to every other part of matter within the solar system, an impulse precisely proportioned to the quantity of matter in the body impelled, and precisely proportioned to the square of that body's distance inversely. If unthinking matter can by its inherent powers exhibit such signs of Wisdom and intelligence, how can we be sure that any Wisdom and intelligence is employed in the government of the world?

It implies that every body, though it has no power to move itself, yet has power to give motion to all other bodies within the solar system.

Sir Isaac Newton thought it inconceivable that inanimate brute matter should, without the mediation of something else, operate upon and affect other matter without mutual contact. This to him *appeared so great an absurdity, that he believed no man, who has in philosophical matters a competent faculty of thinking, can ever fall into it.* Yet we see it is a part of this system of materialism, that matter by an inherent power moves other matter at great distances.

Dr. Priestley tells us that it was objected to his system as an absurdity, that bodies should act where they are not.

In answer to this objection, he acknowledges that there is a considerable difficulty in this case.

He seems to be more apt to perceive absurdities in unions than in disjunctions. The union of soul and body appears to him to be, not a difficulty only, but a great absurdity; but the disjunction of a power from its subject, many millions of miles, appears to be only a considerable difficulty.

Being sensible however of the weight of this difficulty, he endeavours to throw it upon Sir Isaac Newton. *It does not* (he says) *in the | least affect the hypothesis I have adopted concerning matter, any more than that which is commonly received.* And in another place, *I do not say that there is no difficulty in this case, but it is not a difficulty that affects my system more than the common one, and therefore it is no particular business of mine to discuss it.* Illustr. pag. 230 and 232.

One ought to be cautious in charging Sir Isaac Newton with an opinion which he himself believed to be an absurdity, which no man, who in philosophical matters has a competent faculty of thinking, can ever fall into. Yet Dr. Priestley is led to do this in defence of his own system. But let us hear the evidence brought to support this charge.

"According to Sir Isaac Newton's observations, rays of light begin to be reflected from all bodies at a certain distance from their surfaces." This is true, but what follows?

"Yet he considers these rays as reflected by those bodies, by powers inhering in, and properly belonging to those bodies."

This no man who understands Sir Isaac Newton's philosophy will admit.

"So also the gravitation of the earth and of the other planets to the Sun, he considers as produced by a power of attraction properly belonging to the Sun, which is at an immense distance from them."

This likewise is contrary to the whole train of Sir Isaac Newton's philosophy. Sometimes, to avoid circumlocution, he uses expressions which some very inattentive readers have construed as Dr.

Priestley seems to have done, but he gives express and repeated cautions against such a construction.

The weight therefore of this position, That bodies act where they are not, or that an inherent power is disjoined from its subject, lies wholly upon Dr. Priestley's system, and no part of it upon Sir Isaac Newton's.

Dr. Priestley adds, "If Sir Isaac Newton would say, that the impulse, by which light is reflected from any body, and by which planets are driven towards the Sun, is really occasioned by other invisible matter in contact with those bodies which are put in motion, I also am equally at liberty to relieve my hypothesis by the same means."

In this sentence, if I rightly understand it, Sir Isaac Newton is introduced as relieving the hypothesis before unjustly imputed to him by another hypothesis, in order to justify Dr. Priestley in claiming the same privilege.

Sir Isaac Newton says very peremptorily, *Hypotheses non fingo*, and that hypotheses have no place in his philosophy, and therefore to represent him as heaping one hypothesis upon another, is not agreeable to his character and principles.

But I am at a loss to understand how Dr. Priestley's hypothesis is relieved. The hypothesis that is to be relieved is, That a body by an inherent power of attraction essential to it, puts in motion another body at a distance from it. And the hypothesis brought to relieve this hypothesis is, That the motion of the body moved is really occasioned by other invisible matter in contact with it. The second hypothesis, to my understanding, appears to supplant and overturn the first. This, probably, was not what Dr. Priestley meant by relieving it.

However, though he thinks himself at liberty to claim this relief to his hypothesis, he, very prudently, rejects it in the words that follow.

"But the existence of this invisible substance, to the agency of which that great Philosopher ascribes so very much, and which he calls *ether*, has not yet been proved, and is therefore generally supposed not to exist." And he goes on to reason against the existence of this ether, till he reduces it to an absurdity.

Lest any reader less acquainted with the philosophy of Sir Isaac Newton should be led by this sentence to think, that he maintained the existence of such an ether, and that he attempted to explain phaenomena by it, it is proper in justice to Sir Isaac Newton to observe, that he considered Natural Philosophy as a science in which no hypotheses or conjectures are to be admitted till they be legittimately proved by experiments or observations. Conjectures might be useful to lead to experiments by which their truth might be proved or disproved, but ought never to be relied upon. No Philosopher ever kept more strictly to these views of philosophy than he did. His conjectures are always clearly distinguished from his principles, so that no attentive reader can mistake the one for the other. He commonly mentions them in the form of Queries, | but never builds any conclusions, or the solution of any phaenomena upon them. They are not things which he wished to be believed, but things which he wished to be enquired into.

Those who have received them as truths, stamped with the authority of Sir Isaac Newton do him great injustice, and degrade him from that superior rank which he holds, to the level of the common herd of those who, in all ages, have adulterated the oracles of Nature with the impure mixtures of their own fancy.

This great man having observed that many phaenomena of Nature, such as those of magnetism, electricity, the cohesion of hard bodies, corpuscular attraction, chymical affinities, the emission, reflexion, refraction and inflexion of the rays of light, as well as the gravitation of the planets and comets, indicate particular laws of attraction and repulsion; he conceived that possibly all these particular laws might result from some one more general law, which if it could be found, would, as a step in the scale of natural or mechanical causes, bring us nearer to a discovery of the first powerful and intelligent cause of all things.

With this view, he had, early in life, considered whether there might not exist in Nature some very subtile elastick fluid or ether, which pervades the whole material system, and is the cause of those attractions and repulsions which we perceive, particularly of the gravitation of bodies.

But after having had this in contemplation through the course of a long life, he was never able to find any satisfying evidence of the existence of such an ether, far less of its properties by which it might be applied to the solution of phaenomena, and therefore at last left it as a Query, Whether such an ether does really exist or not. And he gives a reason for putting this query, which, though it shews that he had never been able to satisfy himself of the existence of this fluid, does not even express any hope of its being afterward discovered.

Et ne quis gravitatum inter essentiales corporum proprietates me habere existimet, quaestionem unam de ejus cau|sa investiganda subjeci, quaestionem, inquam, quippe qui rem istam experimentis non habeam exploratam. Optice. 2d. Edit 1717.

Dr Priestley adds with regard to this ether, "And indeed if it did exist, I do not see how it could produce the effects ascribed to it. For the particles of this very ether could not impell any substance unless they were impelled themselves in the same direction; and must we provide a still more subtile ether, for the purpose of impelling the particles of the grosser ether? If so, we must do the same for this other ether, and so on, *ad infinitum*, which is absurd."

Sir Isaac Newton was not wont even to put questions that lead to an absurdity. If the effect which Sir Isaac Newton ascribed to his ether, had been to account for the phaenomena of the material system by mere matter and mechanism, without a first intelligent and powerful Mover, which Dr. Priestley's reasoning might lead one to think, his argument to shew the absurdity of such an attempt would have been pertinent. But as Sir Isaac Newton's view was quite of a different kind, the argument does not apply either to him or his ether. No man was ever more sensible of the absurdity of attempting to account for the phaenomena of Nature by mere matter and mechanism than Sir Isaac Newton was. He says, *Elegantissima haec Solis, Planetarum et Cometarum, compages, non nisi consilio et dominis Entis intelligentis et potentis oriri potuit.*

The view that led Sir Isaac Newton to think of his ether has been already mentioned; and it was grand and worthy of the Man. The business of Natural Philosophy is to trace the laws of Nature as far as possible, and from particular laws to rise to those that are more general. The farther we are able to go in this tract, the clearer evidence we shall find that they all ultimately depend upon the will and power of one intelligent Being, who contrived this system by his wisdom, and upholds it by his continual operation.

Dr. Priestley has another relief to his hypothesis in reserve. Illustrations pag 233. "If it be supposed, that no | kind of matter is concerned in producing the above mentioned effects, at a distance from the surface of bodies, but that the Deity himself causes these motions, exerting his influence according to certain laws, am I not at liberty to avail myself of the same assistance"?

Here again the assistance claimed overturns the hypothesis instead of aiding it. That certain effects should be produced by powers inherent in matter, and that no kind of matter should be concerned in producing these effects appears to me, not a difficulty, but an evident contradiction.

In this chapter we have endeavoured to shew, that Dr. Priestley has given no proof that powers of attraction and repulsion are inherent in matter, that his attempt to fix this opinion upon Sir Isaac Newton is unsuccessful; and that the aid he claims, either from subtile matter, or from the immediate operation of the Deity, in order to support his hypothesis, instead of giving any aid to it, subverts and overturns it.

Chap. VI. Consequences of this System.

I shall conclude this Subject by mentioning a Corollary or two from Dr Priestleys Reasoning supposing it to be just, that we may judge of it by its Fruits.

I. If Dr. Priestley, by such an uniform and rigorous adherence, as he professes, to Sir Isaac Newton's rules of philosophizing, has given a just account of the qualities of matter; Sir Isaac Newton must have reasoned very ill in deducing contrary qualities from the same principles.

There is no difference between these two Philosophers, in the experiments or facts from which they reason. Both have the same premises, and the same rules of reasoning; yet their conclusions are contradictory. Sir Isaac Newton concludes that all matter is inert, and that it is passively attracted and repelled by impressed forces; Dr. Priestley that no matter is inert, and that it actively attracts and repels other matter by inherent powers.

If Dr. Priestley reasons justly, Sir Isaac Newton reasons very ill.

It would therefore be of infinite service in Philosophy | to point out the fallacy in reasoning, which imposed upon Sir Isaac Newton and the whole tribe of experimental Philosophers for more than a century, and made them conceive the *inertia* of matter to be a point established beyond the possibility of contradiction.

Such fallacies, in drawing a conclusion from premises, when set up to view, serve as beacons to all future generations, and may prevent their splitting upon the same rock.

It would even give considerable additional force to Dr. Priestley's conclusions, if in addition to his own reasoning from his premises, he had condescended to shew, by what fallacy, so many excellent reasoners had been led from the same premises to contradictory conclusions.

II. If Dr. Priestley's account of the qualities of matter be true, that system of Natural Philosophy which was begun by Gallileo, carried on by succeeding experimental Philosophers, and conceived to be brought by Sir Isaac Newton, to a degree of perfection very honourable to the human Understanding, is overturned to the foundation; and a new one must be raised in its place. Our Mechanicks, our Hydrostaticks and Hydraulicks, our Pneumaticks and Astronomy, are all a delusion.

Experiments, indeed, and facts remain, and are not called in question; but the conclusions deduced from them and the explications given of phaenomena must all be given up as false.

For the whole of this system, which we have been wont to consider as the glory of modern times, leans so much upon the *inertia* of matter, that it cannot possibly stand when that is pulled down. We cannot so much as account for the descent of falling bodies, without supposing matter to be inert.

Sir Isaac Newton's Mathematical Principles of Natural Philosophy are all founded upon the supposition of matter being inert, and if it have no such quality must | be merely chimerical.

A new system of Mathematical Principles of Natural Philosophy must be formed, on the supposition that matter is a substance which has no solidity nor inertia, but solely inherent powers of attraction and repulsion.

It must not be supposed even that it is extended; for in matter that is unsolid and penetrable to matter, extension must be only an accident; it may have more or less, or none at all.

Till such a system be formed of the Mathematical Principles of Natural Philosopy, and applied to the solution of the phaenomena of Nature, we shall have no Natural Philosophy at all, as far as relates to the motion of bodies. The old system, to which we hitherto trusted, being laid in its grave, and the new, being yet an embrio, we must wait till it be brought forth.

X

but to examine a new Notion of Matter, which Dr Priestley has advanced in order to give Aid to his System of Materialism. That Author seems to have been aware that if Matter be what it has been held to be by all Natural Philosophers from the Time of Galileo it is not a proper kind of stuff to make a human Soul. But he thinks that Matter has been unjustly reproached that it has not those Qualities which have been thought inconsistent with the Power of thinking; and that it has inherent active powers, which bring it to a nearer resemblance of what we have been wont to call Spirit. He maintains therefore that Matter is not solid and impenetrable, nor inert, & that it has inherent Active Powers of Attraction & Repulsion, which are essential to it. And the proof of these things he professes to build upon the Rules of Philosophizing laid down by Sir Isaac Newton in the 3d Book of his *Principia*, claiming that this new Doctrine concerning Matter, be tried by that Test & by that onely. I endeavoured to shew that this Doctrine concerning Matter has no solid Foundation neither in the Reasons he has assigned nor in Sir Isaac Newton's Rules of philosophizing, & that if it be true the whole System of Natural Philosophy concerning the Motion of Bodies, laid down by Sir Isaac Newton & carried on by so many great Genius's who have followed him, is so much labour lost being built on a false bottom. For the whole of this System supposes Matter to be inert, and to persevere in its state of Motion or Rest, untill by some impressed force its state is altered. Dr Priestley maintains that no Matter is inert, but has inherent powers of attraction and repulsion. If the last of these positions be true the first must be false, with all that noble Fabrick which has been built upon it.

therefore takeing leave of this Doctrine of Dr Priestley concerning Matter, as a Doctrine which needs itself be better supported than it has been before it can give any Support to Materialism; & takeing it for granted that Matter is that inert & passive Substance which all Natural Philosophy teaches it to be, I shall in this Discourse attempt to draw some consequences from this Doctrine, concerning the Intercourse between the various parts of the natural world that fall within our view, and immaterial Beings or Substances. If the Phenomena of the Material World and the Nature of Matter necessarily require such an Intercourse, Reason requires that we should admit it, however contrary it may be to groundless prejudices.

All the Matter that comes within our view, has been very properly divided into three Classes, or Kingdoms as they have been called. to wit Inanimate Matter, Vegetables, and Animals. Although these different kingdoms border so nearly that we are sometimes at a loss whether some particular body belongs to one or to another kingdom. Yet this is owing to the imperfection of our Senses or to our Ignorance ⟨of⟩ the body which is the Subject of Dispute. For the distinction of these different Kingdoms is very clear in our Conception, & the error is onely in the Application of it to particulars. In a piece of inanimate Matter there is nothing which has a relation to the whole unless it be figure or dimension or Use or some such external denomination. In vegetables & animals there is life which removes when the whole is taken to pieces & is not found in any of the parts nor can be restored by the Art of Man. The most curious composition of inanimate matter cannot draw nourishment assimilate it so as to become a part of its own Substance extend its dimensions, cure its diseases, heal its wounds & propagate its kind

By these and such like Qualities, inanimate Matter is distinguished from that wh⟨ich⟩ hath life either as a vegetable or animal. The distinction between Vegetables & Animals is no less real nor less distinctly conceived. Animal Life implies some kind or degree of thought be it ever so small. The least degree of Sensation suppose it onely the feeling of a small degree of pain or pleasure is called animal Life, which we see no reason to ascribe to vegetables. The distinction of the different classes of Material beings being thus asscertained I aprehend that from the inactivity of Matter compared with the Phenomena of the Material World we may draw these two conclusions. I That all the inanimate Matter that falls within our view is constantly acted upon by something immaterial. 2ly That both vegetables and Animals are United to something immaterial, by such a Union as we conceive between Soul and Body,

Part Three: Materialism 219

which Union continues while the Animal or Vegetable is alive, & is dissolved when it dies.

To illustrate these two Conclusions is the Intension of this Discourse

To prevent missunderstanding of the Term *immaterial* I would observe that the terms material & immaterial being contradictorily opposed must according to the principles of Logick include between them every thing that exists. For every thing either is matter or is not matter which is all that I mean by immaterial. Some may be apt to think that what is immaterial must be endowed with thought and Intelligence. But this I take to be a mere Prejudice, like that of the Inhabitants of the Canary Islands when they were discovered by the Spaniards in the year 1405; who having had no Intercourse with the rest of Mankind from the destruction of the Carthaginian Empire, believed that there was no land on the face of the Earth but their own Islands nor any Men but themselves. Although the Beings we are acquainted with be either material, or thinking & Intelligent, there may, for what we know be many Orders of Beings which are neither material, nor thinking & intelligent. But every being that is not material must be immaterial

These things being understood I say that all the inanimate Matter we are acquainted with is constantly acted upon in various ways by something that is immaterial. This is a necessary Consequence of the Phenomena of the Cohesion of the parts of hard bodies, of the Phenomena of Corpuscular Attraction, the Phenomena of chemical Affinities, of Elasticity both in hard & in fluid Bodies⟨,⟩ of Magnetism, Electricity and Gravitation. In all these we see active Power exerted, in producing Motion in accelerating or retarding it, or altering its direction. These active Powers cannot be in dead Matter. They are as Sir Isaac Newton very properly calls them impressed forces. Matter is the passive Object on which the impression is made, but what is it that actively gives this Impression. It cannot be matter, for to suppose Matter wholly inactive and active at the same time is a contradiction. To begin Motion which did not exist before to accelerate or retard it or alter its direction are Effects which require an active caus. In mere matter there is no activity, & therefore it cannot be the Cause of such Effects

If it should be objected to this Reasoning, That though Matter cannot begin Motion, yet Motion may be begun in it by the impulse of other Matter already in Motion, & that Sir Isaac Newton thought it not impossible that Gravitation & the other powers above mentioned might be produced by some Elastick Fluid so subtile as to pervade all Bodies & to be indiscernible by our Senses. |

[A verso] In answer to this Objection, it is admitted that a body may be put in Motion by the impulses of another Body already in Motion, which must lose as much Motion as it communicates. But from what cause did the impelling Body get that Motion which it communicates? Shall we say that it got this Motion by the impulse of a third and that by the impulse of a fourth and so backward *in infinitum* without any beginning or first cause of this communicated Motion? This I think all Men acknowledge to be absurd. This Motion therefore, through Whatever Number of Bodies it may have been communicated & have passed must have had a beginning, and a Cause, and it is evident that Matter could not be the Cause of the first Motion given to it. The cause of this first Motion must have been something immaterial.

As to Sir Isaac Newton's Ether it is to be observed that he never affirmed the Existence of Such an Ether. He onely put it as a Question whether there existed any such Ether and he tells us that the Reason of his putting it was, lest any one should imagine that he took Gravitation to be a quality essential to Matter. He thought it not impossible that Gravitation might have a physical Cause, which, could it have been discovered would have added a link to the chain of physical Causes, but he was perfectly aware that this chain of physical Causes must have a beginning and that its whole weight must hang upon some efficient Cause which is immaterial and active. |

[B1r] Nor do I think that any considerable Objection against the Conclusion I have advanced can be drawn from the Hypothesis of Boscowitsch. That ingenious Author supposes that Matter is composed of Mathematical points which have alternate Spheres of attraction and Repulsion. By which he thinks all the Phenomena of the material System may be accounted for.

This Hypothesis indeed seems to leave no such thing as matter in the Universe, unless we give the Name of Matter to that which hath neither length breadth nor thinkness. But to pass this let us suppose these points to be Matter. The alternate Attractions &

Repulsions of these Points imply an Active Power which cannot be inherent in inert Matter It must be a Force impressed by something that is not Material

If it should be objected that he who said *Let there be Light & there was Light* may have said Let Matter gravitate & the Effect must follow, so that no other Cause of Gravitation is necessary, but the will of the Supreme Being, establishing it as a Law of Nature. To this I answer That will without Power can produce no effect. He who said *Let there be Light* at the same Time exerted his infinite Power in creating Light and if in like manner the Gravitation of Matter and the other primary Laws of Nature which govern the Material System be the immediate Effect of Almighty Power without the Intervention of any Subordinate Being this is no Objection to what I plead for, but perfectly agrees with it, to wit That the Gravitation of Matter and the other Laws of Nature mentioned are forces impressed ⟨on⟩ Matter by some Being that is Not Material. Whether it be by the supreme Being immediately or by some Subordinate Being or Beings appointed by him for that purpose, is I apprehend quite beyond the Reach of human Knowledge We know so little how we are enabled to act upon Matter our Selves that it is vain to think we shall be able to comprehend how other immaterial beings may act upon it.

These forces appear to be constantly impressed without any variation excepting in the case of Miracles. We commonly consider a Miracle as an exertion of Power in some invisible Being, but it may be onely a temporary suspension of an Exertion which at other times is constant

The constancy of the Laws of Nature and of these Exertions of invisible Beings by which they are executed is the foundation of all our Knowledge of the Material System, and of all our Power over it. From the Regularity of the Phenomena Rational Creatures may conclude by Induction the Laws by which they are regulated and by the knowledge of those Laws we are enabled, both to account for other Phenomena, and to forsee the Effects of the various Operations upon Matter that are within the Compass of human Power.

To observe and arrange the Phenomena of the Material System, so as to discover the Laws by which they are Regulated, & from the Knowledge thus acquired of the Laws of Nature, to account for other Phenomena of Nature, and to devise such applications of the Laws of Nature as may enlarge our Power over the Material System, & increase human happiness and comfort, is the Aim of Natural Philosophy, and the End in which it rests.

But the Rational Mind is naturally led to go farther, and to consider how those Laws of Nature that regard the material System have been Contrived and how they are executed. In the contrivance of them there appears such Intelligence Wisdom Goodness, such Uniformity & Consistency, as leads us to ascribe this Contrivance to an infinitely powerfull Intelligent & Good Being on whom we and all things intirely depend. As to the Execution of the Laws of Nature that obtain in the Material System we know very little, but from what we know of the Inertness and Inactivity of Mat⟨t⟩er and from the Active Forces we see every where employed in producing Phenomena, we may I think certainly conclude that these Forces are exerted upon Matter, either by the immediate Agency of the immaterial Cause of all Things or by the Agency of subordinate immaterial Causes or Instruments which he has appointed for that purpose

What the Number, and the Powers of those Subordinate Agents may be if such there are is hid from Men. Perhaps that Man who has always been prone to Idolatry ⟨will⟩ be under no Temptation to give that Homage to them which is due onely to their end our Creator. |

[B1v] I consider it therefore as beyond the reach of human Knowledge, whether the Forces that act upon Matter in this our System be impressed by the immediate Power of the Supreme Being or by Subordinate Agents appointed by him Revelation every where ascribes the Phenomena of Nature to the first Cause, to whom they ought ultimately to be referred, whether subordinate Causes be employed by him for this purpose or not.

Thus we see that the Inertia or Inactivity of Matter compared with the Phenomena of the inanimate part of our System leads us to conclude, that all the inanimate Matter we are acquainted with is constantly acted upon, either by the first Cause or by subordinate immaterial Causes, so as to produce the Laws of Gravitation, Magnetism, Electricity Cohesion corpuscular Attraction & the chemical Affinities, and perhaps other Laws of Nature yet undiscovered. The Argument is short and simple – In all these Laws of Nature

Active power is constantly exerted; There is no such Power in Matter; therefore, it must be in some Being that is immaterial. By these impressions upon Matter it is made a fit Instrument for all the purposes of animal Life. And by these Impressions on Matter, Man who is endowed with Reason is led to perceive his Connection with an immaterial & invisible World.

Having thus considered the Intercourse between the Inanimate Kingdom of Nature and immaterial Beings Let us next attend to the Intercourse which our Reason discovers to be between the animated Kingdoms of Nature I mean Vegetables & Animals, & immaterial Beings. All that is the Object of our Senses both in Vegetables and Animals is inert Matter, subject to the Laws of Motion, subject to all the Impressions of Gravitation, Cohesion, the corpuscular Affinities, and the other Laws of Nature that act upon inanimate Bodies. Every part of an Animal or a vegetable was once inanimated Matter and when the Animal or Vegetable dies, becomes inanimated Matter again, fit to be assimilated into the Substance of some other Animal or Vegetable. As all the concrete Bodies we know whether animated or inanimate are compounded of various Elements, which adhere by their mutual affinities, some of these Elements in their Elementary state being Volatile, As all those called Gases, other⟨s⟩ fixed in all degrees of heat we are able to produce as Earths and inflammable Bodies. By various combinations of these Elements all animated as well as concrete Bodies are produced and from their combination their forms and Qualities are derived. There are no doubt combinations in vegetable bodies, and others in animal Bodies that are peculiar to their several Kingdoms. But the Elements are the same in all and have the same properties. When the Vegetable or Animal dies those Combinations which were peculiar to the Animal or vegetable Life are either more quickly or more slowly dissolved by the surrounding Elements, by Fermentation, Putrefaction, combustion or other processes of Nature, so as to leave no vestige that they once made a part of an animated body & belonged to the Animal or Vegetable Kingdoms.

Is there then no other difference (it may be asked) between animated and inanimate Matter but this, That while both are compounded of the same dead Elements and while Many combinations of those Elements are common to both, the Animated Body has some peculiar combinations of those Elements which the inanimate has not? I answer to this Question that it is by no means so. The Animated Matter has united to it a Principle of Life, which pervades the whole Animal or Vegetable, and Unites it into one Being, which has various Members & various active Functions relative to the whole. As soon as this vital Principle departs, all the material Elements remain but are dead & inactive, Even those combinations of Elements which were peculiar to its animated state may remain for a long time, so as it may be known that they once belonged to an animated Body, but they retain no symptom of the life that once animated them

We are therefore led to enquire what this Principle of Life is which distinguishes animated from inanimate Matter.

If it be asked what Life is, we are at a loss to give a distinct Answer. We know it onely by its Effects, & we know that those Effects must have a Cause. But the Cause is no object of our Senses. We know not when life begins in any Being that hath Life, whether Vegetable or Animal. Nor do we know precisely the Moment it ends. It may in some individuals ly dormant for a long time, without shewing to our Senses the Active Functions which at other times it exhibits. Thus the seed of a plant may for years have all the appearance of ⟨an⟩ inanimate body; untill a certain degree of heat and moisture stimulate its dormant Powers then its Life is plainly discovered. But if the Principle of Life | be gone, no degree of heat or Moisture has any other Effect but to hasten its dissolution. Perhaps this seed even in its dormant state had some Activity, by which its organization was preserved from the impressions of the Surrounding Elements, & by which perhaps that degree of heat was preserved which was necessary to its Life.

Every thing that hath Life whether vegetable or Animal hath an Unity which does not belong to dead Matter. A piece of dead Matter is from its Nature made up of parts and divisible into parts which may not onely be separately conceived, but really have a separate existence so that one part may be annihilate without affecting the remaining parts. But to suppose a Life made up of parts of Life or divisible into parts is absurd. We cannot affix a meaning to a part of Life. The Principle of Life therefore in every Being that has Life is one & indivisible.

A Position so evident to our Reason as this is, cannot be overturned by a fact which is well known, that some Animals and many vegetables being cut into two or more pieces each piece will grow into an Animal or Vegetable of the same kind. This fact may easily be reconciled to the principle of reason we have advanced. The ways by which animals and vegetables produce their kind are various and all equally mysterious & incomprehensible to human understanding. In some animals there is one womb destined for this mysterious operation in which the *Fœtus* is formed, and is either excluded in the form of an animal of the species, or in the form of an Egg which may long remain without the signs of Life that are discernible by our Senses, and may in some require Incubation in others not. In other animals and vegetables there are many organs in various parts of the body fitted for the production of a Fetus, which in some grows like a branch from the trunk till in its maturity it drops off from the parent a perfect animal of the Species, in others drops in the form of a seed or egg which in time & perhaps after various transmutations, becomes like to its Parent. It is therefore probable that every part of an Animal or Vegetable which being cut off forms an animal or vegetable of the same kind had before its separation from the Parent contained a living Fœtus which had its own Life before separation, and carried off no part of the Life of the Parent.

An Animal or vegetable which to our Senses appears one & single may contain within it many animated *Fœtus's* each of which has its own Life, and is capable of growing to its perfection when disjoyned from the Parent. This is known to be the Case of a pregnant Mother or Dam, & of a Tree in seed. The multiplying a Worm or Insect therefore by cutting it into pieces is onely a Phænomenon similar to the Cæsarean Operation which has been performed in the human Species. Whether in the Case of the Worm, the Parent dies, and the *Fetus's* onely survive, is beyond our power to discern, though it seems most probable.

I know nothing therefore in the Phenomena of Nature that hinders us to conclude what our Reason plainly dictates that every living thing whether vegetable or animal has an Unity & individuality which belongs not to Matter

There are other Properties of Life more obvious to our Senses, by which it is distinguished from Matter. One very remarkable is that while Life continues it preserves the Organization of the living Being. This Organization as was already observed consists of a

Combination peculiar to that Species of the Elements of common Matter How this Organization is at first formed is beyond human Understanding. Though the Elements of which it is made are every where at our hand, & Man by his Art and industry has learned to form an infinite Variety of Combinations of those Elements, yet no Art of Man could ever combine them in such a manner as to form the Flesh of an Animal, the Timber of a Tree, or so much as a single Fibre of a Muscle. But in whatever Manner this Organization was formed at first, it appears to be preserved by the Life | of the Being to which it belongs. That Life while it continues protects the Organization from the impressions of the surrounding Elements. But Life is no sooner gone, than the Air and Heat and Moisture which were Salutary while Life continued, now attack the organised but dead Body, on all hands and continue to prey upon it, till sooner or latter they destroy its Organization & reduce it into its original Elements, or turn it into combinations of a quite different Nature. Life not onely protect⟨s⟩ living Beings from that corruption which Air and Moisture and Heat produce as soon as they die; It seems likewise to protect them from the power of Digestion in the Stomachs of Animals

It is a Phenomenon altogether unaccountable from Mechanical Causes, that the Gastrick Juice or whatever it is that has the power of Digestion, though elaborated and retained in the Stomach, never exerts its power upon the Stomach itself, when alive, while it digests and dissolves every kind of animal and vegetable food The stomach that contains it is as fit to be digested when dead, as any of those kinds which were digested in it without hurt when alive. We are told that in sudden Deaths this Gastrick Juice often attacks the stomach itself as soon as the Principle of Life has deserted it. And it is well known that some Animals can live without hurt in the Stomach & intestines of others but as soon as they die they are digested. Some Animals, as Crabs and other Shell Animals are said to change their Stomach annually or periodically, a new Stomach growing gradually around the old. As soon as the new Stomach is perfect the principle of Life deserts the old, it becomes dead Matter, and then is digested like other food. From these Phenomena and others that might be named it seems evident that there is some unaccountable Property in the Principle of Life by which it preserves and protects the Organization of its

Part Three: Materialism

own Body, from those Powers of Dissolution which as soon as that Protection is withdrawn, attack and gradually dissolve that Organization.

In general there are many Functions peculiar to living Beings which cannot be the Effects of Inert Matter. The drawing Nourishment from dead Matter the digesting it and assimilating it to their own Substance & their Peculiar Organization, the Distributing it to the various parts of the body so as to produce their increase and growth, The Glandular Secretions from the common Aliment of various Fluids proper to the Species, the throwing off the fecal part by Organs destined for that purpose. *and above all the producing a progeny after their kind in a manner proper to that kind, and so various in different kinds.* The Powers of expelling what is noxious, of healing their diseases, cuatrizing wounds and in some cases of regenerating lost Members. These Functions of Life seem to me plainly to indicate an intercourse between Matter and immaterial Beings of a quite different Nature from that constant and uniform Action of immaterial Beings upon the inanimated kindom of Nature, which we before endeavoured to prove. Every vegetable and animal seems to be a System of Matter united with and animated by some active immaterial Being which makes it one Individual, protects its organization from the impression of the surrounding elements and performs occasionally as need requires those vital Functions in its proper Body which have been Mentioned, in a manner inscrutable by human knowledge & inimitable by human Art.

Much is said by modern Physiologists of *Stimuli* and of Irritation as the cause of the Animal & vegetable functions in living Beings. The seed of a Plant is stimulated by the application of heat & moisture, & by that Stimulus is made to put forth its vegetable Powers. The Heart & arteries are stimulated by the Blood & therby perform their pulsations. In the same sense I apprehend we may say that Iron is stimulated by the Magnet & made to approach it or to flee from it, and that the Sun stimulates the Planets & Comets and produces Gravitation.

If by this manner of speaking no more be meant but to express an Effect evident to Sense while we confess our Ignorance of the Cause of that Effect, I have no objection to it. We may use the words Stimulus and irritation as we use the Words Gravitation and Attraction, not to express an Efficient Cause, but to express an Effect of which the Efficient is beyond our Knowledge. But if any should be so weak as to imagine that by this Language of *Stimuli* and Irritation any of the vegetable or animal Functions is accounted for, without | the Intercourse & Efficiency of some immaterial Cause, they deceive themselves and mistake ambiguous words for Efficient Causes. We know that by the Laws of Motion which are the consequences of the Inertia of Matter one Body may be moved by the impulse of another. But the Motion of the Body impelled must both in its Quantity and direction be just what the impelling body loses. If therefore a Stimulus be at rest it cannot be the cause of Motion in a Body at rest, & if in Motion it can onely communicate a part or the whole of its own Motion to another body. Whatever Motion beyond this follows the Application of the Stimulus must have another Efficient Cause When heat and Moisture are applied to a living seed, its vegetable powers which before seemed dormant are excited to action, it pushes its roots downward and its stem upwards. But are these motions communicated by the impulse of heat and Moisture so one body is put in motion by the impulse of another. It is impossible The Motions of the seed are evidently produced by the Living Principle within it which has the power of producing these Motions, for when it is dead no such Motion is produced by heat & moisture

When a Ventricle of the heart is said to be irritated into contraction by the stimulus of the blood distending it. If it be meant that it is contracted onely by the Elasticity of its muscular Fibres. This cannot be true, for in that case it could contract with a Force no greater than that with which it is dilated. But it is evident that the Force of its contraction must be as much greater than that of its dilatation as the Momentum of the Blood propelled by the Heart and Arteries is greater than its Momentum in its return by the Veins. This additional Force which it exerts in its contraction, can have no mechanical cause, it is inherent in the living Animal. Not to mention that elasticity even in dead Matter implies the constant Action of some immaterial cause as has been before observed.

Thus I think it appears that those Authors who ascribe vegetable or animal Motions to the Irritation of Stimuli, if they understand themselves, cannot mean by this Language to assign the Efficient Cause of such Motions, but onely mean to express the Effect and

Part Three: Materialism

the Circumstances that accompany that Effect while they acknowledge the Efficient Cause to be latent. There is evidently an Active Power exerted both by Animals and Vegetables. Active Power implies an active agent, and inert Matter can be no such Agent

From all that has been said it seems reasonable to conclude that as inanimate Matter is constantly acted upon by immaterial Beings so as to produce its Gravitation Cohesion and the various Corpuscular Affinities and Attractions which Natural Philosophy has discovered, so Animals and Vegetables are animated by some immaterial Being which is the Efficient Cause of their Animal & Vegetable Functions while they live & which is separated from them when they die.

I am not much moved by the Questions which this Doctrine may give occasion to. It may be asked to what Order of Beings we must refer those immaterial Agents that act upon inanimate Matter, & those by which Animals and Vegetables are Animated? Are they thinking intelligent Beings or not are they mortal or immortal? What becomes of the Soul of Plants & A⟨n⟩imals when they die. To these and such Questions I can answer onely by confessing my Ignorance. Every Advance we make in Knowledge opens to our View Regions unknown. Our Ignorance of these ought to be no objection to what our Reason has discovered. What I have endeavoured to shew of those Beings which constantly act upon inanimate Matter, & those which animate Vegetables and Animals, is that they are not Matter. They are no doubt by their Maker endowed with the Powers necessary for discharging the Offices to which his wise Providence has destined them. We see no evidence of thought in Vegetables, though they plainly discover Activity and self Motion In Animals we perceive the signs of Sensation and of various kinds of thought. In some tribes, the thinking Principle seems more perfect than in others, though in all brute Animals far inferior to what Man possesses. For it is the Superiority of his mental Powers onely that gives him Dominion over all the other Animals that inhabit the Earth.

Upon the whole if there be such intercourse between the Matter of this our System & immaterial Beings, we may see that what our Senses discover of the World in which we live is the most inconsiderable part of it. The Earth the Sea and Air are full of invisible & immaterial Beings which either act upon it in the various ways that we call Laws of Nature, or are united to and animat⟨e⟩ certain Systems of Matter which we call | brute Animals and Vegetables. So that as we are taught to believe that there are many Orders of invisible Beings superior to Man, our Reason discovers that there are also many Orders of invisible Beings inferior to him, & destined to meaner purposes in the Creation of God. Thus the Philosophy of Matter naturally leads to the Philosophy of immaterial Being.

To conclude if the meannest Animals and even Vegetables be endowed with an immaterial Principle there can remain no doubt of the Existence of such a Principle in Man. That the thinking Principle in our Composition is immaterial appears agreable to the whole Analogy of Nature. Those who have treated of the imateriality of the human Soul seem to have overlooked this Analogy, or rather have thought it dangerous to admit it. Yet it would be difficult to shew that the Arguments which prove an immaterial Principle in Man, do not extend to all animals and even to vegetables, since we allow to them Life & Activity which cannot be in inactive and inert Matter nor do I see any danger in admitting this Consequence. Man is distinguished from his fellow Animals by a rational & Moral Nature, which is the Image of his Maker, And this is a sufficient foundation for those Prerogatives we justly ascribe to him.

In all that has been said I have taken it for granted that Matter is Inert and meerely passive, because this is a fundamental Principle in Natural Philosophy, & because in the Discourses formerly alluded to I endeavoured to answer what Dr Priestley has said against it. It is indeed a very strange Contradiction in that celebrated Author, while he denies the Inertia of Matter to admit the Truth of the Laws of Motion which are grounded upon the Inertia of Matter, & to pretend to ground his Doctrine upon the Laws of Philosophizing of Sir Isaac Newton whose Philosophy in every page rests upon that foundation.

Before I make an end of this Discourse it is proper to take Notice of an Opinion or Conjecture of Mr Locke of which though he was no Materialist, those of that Perswasion have availed themselves Much It is that though Matter be of itself incapable of thought yet by the Power of the Almighty it may be endued with the Power of thinking. Far be it from us to limit the Power of the Almighty in any thing that is possible to be done. If Mr Locke had believed that the essential Qualities of Matter and of thought are such as that both cannot subsist in the same Subject, he certainly would have changed his Opinion. Since Mr Locke wrote two attempts have been made to prove this, by which the imateriality of thinking Substances is put in a clearer Light than it was before. Dr Samuel Clarke one of the greatest Metaphysicians of this Age, has endeavoured to shew that the divisibility of Matter is inconsistent with its being the Subject of Thought. Matter is made up of parts each of which is a distin⟨c⟩t Substance & has its own inherent Qualities. Every Quality in the whole is compounded of parts of the same Quality inherent in some or in all the parts of the compound substance. Every part being a distinct Substance whatever is inherent in one part cannot be the same that is inherent in another part. External Denominations such as Figure or Place or Use may be common to the Whole, but no inherent Quality can. Thinking is an inherent Quality in its proper Subject, the thinking Being. To suppose one part of a thought to be in one Substance and another part of the same Thought in another Substance appears to be absurd. The Subject of Thought therefore must be one individual Substance. But no Matter nor part of Matter is one individual Substance & therefore Matter cannot be the Subject of Thought.

This Argument was very fully examined by Mr Collins the most Acute of those who took to themselves the name of Freethinkers. His Objections & Dr Clarks Answers (which appear to me very satisfying) are extant in several Replies & Duplies. of which therefore I shall say nothing more

Another Attempt to shew that Matter cannot be the Subject of Thought was made by the celebrated Euler. in an Essay on that Subject. His Argument is taken from the Inertia of Matter which implies its being wholly passive & unable to alter its state either of Motion or of Rest. As this cannot be the case of a thinking Being, it seems as evident that Matter cannot be the subject of Thought, as that the Same | Being cannot be totally inactive and active at the same Time.

If it should be said that Inertia is not essential to Matter, & that there may be Matter without this Quality. I think we have strong Reasons to believe the Contrary. For this quality is not onely found in all Matter on which we can make Experiments, & is found to be precisely in proportion to the Quantity of Matter, without ever being increased or diminished, and is found to be the same in

Animals and Vegetables as in inanimate Matter; but it seems very difficult or rather impossible to conceive Matter deprived of it. For can any Man conceive that Matter at rest can be put in Motion without any Force? Or that a greater Force is not required to produce a greater quantity of Motion than to produce a less? This surely would be to suppose an Effect without any adequate Cause. But the Inertia of Matter is the onely cause why Force is necessary to move it when at rest or to change its direction or velocity when in Motion. This Inertia is the exact Measure of the Force required in any change of Motion & the onely Resistance which that Force has to overcome.

Xa

I shall take a general View of the Material World that falls within our Notice, and endeavour to shew that in all its parts Animal Vegetable & inanimate, it has such connection with immaterial Substances, as renders an Union of an immaterial thinking principle with a human Body analogous to all that we see in the general course of Nature.

Let us first consider the inanimated parts of Matter. In these we perceive nothing that can be called Life or Death, no organization consisting of various Members connected together, having fluids circulating through a whole System, by which it is capable of being fed and nourished, of growing up to its perfect state & propagating its kind. By the mechanical Arts, parcels of inanimate Matter may be divided in to parts, or put together in various forms or figures; so as to produce tools utensils & machines for different Uses. But no Mechanism can add any new principle to the Matter it compounds. When the most curious Machine is taken to pieces there is no principle or part of the composition lost by this division nor is there any thing added by the composition. It is otherwise with vegetables and animals, by taking them to pieces something is lost which we call Life. Which is not restored by putting the parts together again. There is therefore a Specifick Difference between animated Matter and that which is inanimate. Although there are such different degrees of Animation that we may be often at a loss to discern the line that distinguishes these two Species.

But in inanimate Matter, though we see nothing like Life or Death yet

XI

Some time ago I had the Honour to read to this Society Some Observations on the System of Materialism advanced and defended by the Revd Dr Priestley. It was not the Intention of those Observations, to discuss the Question at large Whether the Soul be a Material Substance or not *but onely* to consider the Aid which that Author had endeavoured to give to Materialism, by giving a new Conception of the Nature of Body or Matter. Dr Priestley seems to have been aware that if Matter be really that inert and passive Substance which all Natural Philosophers from Galileo downwards supposes it to be, that it is not a proper kind of stuff to make a Soul. But this prejudice he thinks will be removed, if we find that Matter has been unjustly reproached, that it really has not those qualities which have been deemed inconsistent with the power of thinking. And that it has active powers which bring it to a nearer resemblance of what we have been used to call Spiritual Substance Therefore he endeavours to prove that Matter is neither solid and impenetrable nor inert, and that it has inherent powers of Attraction and Repulsion. I endeavoured to show that these new Notions concerning the Nature of Matter have no solid foundation in Reason or Experience from all that Dr Priestley has said in support of them & therefore that no Aid to Materialism can be derived from them it must still stand upon the old Ground, and that if they were true, they would from the foundation overturn the whole System of Natural Philosophy relating to the Motion of bodies. In this Discourse, setting aside the peculiarities of Doctor Priestley I intend to consider at large whether we have ground to think the soul to be a Material Substance or not?

This Question is the more difficult to be determined, because we know so little about Substances of any kind Our Senses inform us of some qualities of Matter; & our consciousness informs us of some Operations of the Mind; but the subject of those qualities, & the Agent of those Operations, are behind the Scene, & hid from us; Substance is neither an Object of Sense nor of Consciousness.

That Qualities and Attributes of every kind must have a Subject, & that Operations must have An Agent is evident. These are necessary truths, and implied in the Structure of all Languages But of this Subject, and of this Agent, we have onely a Relative & therefore an obscure Conception We cannot immediately compare them, so as to discern whether they be the same or different. And as it is onely by means of their Attributes that we get the conception of them, so it is onely by means of these that we can judge of their identity or diversity

There seem therefore to be onely two ways by which we can judge whether the Subject of thought and the Subject of sensible qualities be the same or different. One way is by comparing in general the Attributes of Mind with those of Body. If these be very similar in kind & in dignity, there is great probability that their Subject is of the same kind. But if on the other hand there be no similitude between the Attributes of Mind & those of Body, if they be very different in point of Dignity & if there appears no necessary connection between one and the other, there is great probability that the Subjects are as different in kind & in Rank as the Attributes are.

When we observe the Variety which is to be found in the Works of God that fall within our Notice, we see no reason to think that he has made onely one kind of Substance. It is more agreable to all that we see to think that he has made various kinds different in Rank and Dignity the lower Ranks subservient ⟨to⟩ the higher, as we see that dead Matter is subservient to vegetables the vegetables to animals, & the brute animals to Man.

There is therefore no presumption from what we see of the general Course of Nature against Soul and | Body being different Substances but rather the contrary. | Nor does there appear to me any Reason to conclude, that every Substance which is not material, must be a thinking Substance It is true that Matter is the onely kind of Substance which we perceive by our external Senses. And that sentient & thinking beings are the onely kind we know by consciousness & Reflection but it would be rash & presumpt⟨u⟩ous from this to conclude that all the variety of Gods Creation must be comprehended under one or the other of these two kind. There appears to be an immense interval between passive Matter on the one hand and the lowest degree ⟨of⟩ active intelligent Being on the other. We see no such Chasm in the Works of God. This interval therefore may for what we know, and most probably is filled up by beings of an intermediate Nature, which are neither material on the one

Part Three: Materialism 235

hand nor sentient & intelligent in the other. And I hope it will appear from what follows in this Discourse that however ignorant we may be of the Nature and Attributes of such Beings we have strong Reasons to believe their Existence

I shall therefore before entering on those two points before mentioned, by which we may judge whether the thinking principle in Man be Material or immaterial |

If it should be thought that the Union of Soul & Body in one person is a presumption of their being one and the same Substance. Let it be observed that in the Operations of Nature we see many instances of the Union of different Substances into one concrete, in which the Nature and Manner of the Union and the effects of it are as far beyond the comprehension of the human faculties as the Union of Soul & Body. By the external Senses and by an infinite variety of Experiments, Men have been enabled to discover many wonders in the Composition of concrete Bodies. The ingredients into which they can be resolved by the Art of Chemistry are such as no human Wisdom or Sagacity could have conjectured. The manner in which these ingredients are united by Nature and the Effects of that Union in the properties of the body compounded of them are as far beyond the comprehension of the human Faculties.

Why should it be thought incredible that Man should be compounded of two different Substances of Soul and Body united in a Manner beyond our Comprehension, when every thing we see and handle and taste is compounded of different Substances united in a manner no less incomprehensible to human Understanding

What is there in Nature to appearance more simple and uncompounded than pure Water. In all Ages from the beginning of the World, it had been considered as an Element, without any Composition from different ingredients. When examined by the Eye aided by the best Microscopes, it shews no sign of composition, every part appears similar to every other part and to the Whole. This is the Case not onely while it is in the form of Water, but when in various degrees of cold and heat it is turned in to ice or Steam. Yet this Substance which from the beginning of the World has been deemed the most simple and uncompounded production of Nature, has lately been found to be a composition of two kinds of elastic Air extreamly dissimilar to it and to each other. But how these ingredients are united, and how the appearance and the properties of the compound come to be so very different from those of either of the ingredients, exceeds the comprehension of the human Understanding.

The Science of Chemistry presents to our view innumerable Instances of the Union of two or more different Substances into one Compound, in which we cannot perceive the least appearance of composition. And although we can examine by that Art the different Ingredients both separately and in their United State, no human Wisdom has been able to discover the manner of their Union, or how it produces those properties which we find to belong to the compound. This remains a Mystery beyond the reach of human Understanding.

The Principle which we call Life in Man and in other Animals presents to a contemplative Mind many things very Mysterious. We perceive its Effects while it is united with the Body, by the change produced when it is removed by Death. By this we are authorised to ascribe to it not onely the Sensations, the Animal Motions and all the Mental Powers the living Animal possessed, but even the Organization of the Body. Every Animal has a Body very curiously organized. How this Organization began, or what part the Principle of Life had in its first formation, is perhaps beyond the reach of our knowledge. But we see evidently the preservation of this Organization, the Nourishment & Growth of the Animal till its parts and functions are perfected, depends upon the Principle of Life. For we see that at whatever period of the animals existence Life departs, that organization, which before seemed to be proof against the Surrounding Elements, is now no longer able to resist their impression. It becomes a prey to corruption its organization is dissolved and all the parts of its composition mix with their kindred Elements. From this it appears evident that Life, whatever it be, was united to every part at least to every organized part of the Body, and was that which embalmed it, as it were, and preserved it from Corruption.

If from this general View of the Effects of Life and Death it were made a Question. Whether this Principle of Life be an invisible Substance united with the body during Life & separated from | it at Death, or whether all the Phenomena of Life be the result of a particular Organization of the Body or of some part of it & Death the breaking of some main Spring or deranging some Essential part of the Machinery. I see no reasonable presumption against the first of these Hypotheses more than against the last, but rather the contrary.

When we preceive the same Body put on such different Appearances as that between a dead Body of an Animal and a living. It may be a question, Whether the change is owing solely to a different configuration or arrangement of the parts of the same Substance, or whether the change is owing to some invisible substance which was joyned with the body in one state and disjoyned in the other. It is no presumption against the last of these Suppositions that nothing discernable by our Senses is disjoyned, because we know that there are Substances which are not perceivable by our Senses.

We see that Water has very different appearances in different degrees of Heat. In one degree it is hard and Solid Ice, in another it is fluid and its parts seem to have no cohesion, and in a still greater degree of heat it is steam, an invisible elastick fluid many hundred times lighter than Water or Ice, whose parts instead of cohering strongly as they do in Ice repell each other with great force. The Question might here be put whether these different appearances of the same Body are onely different Configurations of one uncompounded Substance? or if that which we call heat be a distinct Substance which being united with Water in various degrees produces such different appearances? The last supposition is now generally thought more probable by Philosophers, though to our Senses nothing appears to be taken from Steam when reduced to water, or from Water when reduced to Ice.

The late Improvements in the Science of Chemistry, as they make a valuable addition to the Fabrick of human knowledge in other Respects, so I think them valuable in this respect that they teach us more of our Ignorance, with regard to Unions and Compositions in things which our natural prejudices would lead us to believe to be the most simple and uncompounded. They teach us by well attested instances, that Substances which to all common Appearance, seem to be most simple and homogeneous in all their parts, may notwithstanding be compounded of various Substances very dissimilar to each other and to the compound. They teach us, that when we have discovered such a composition by certain and infallible proof, it may still be beyond the reach of our Understanding to know how or by what tie the ingredients which are said to have an affinity are united, and how the properties of the compound result from that Union. They teach us that in such compositions, different degrees of the same ingredient may produce such different qualities in the compound. as to appear not to differ in degree but to be contrary to each other

Thus Water united with one degree of heat is a hard and solid body whose parts adhere together with great force. With a greater degree of heat it becomes a fluid, whose parts have no sensible cohesion, and with a still greater degree of heat, it becomes Steam, whose parts repell each other with great force. Nor are we able to discover how, heat produces any of these Effects, or how its various degrees produce such contrary Effects.

Farther I think that both Chemistry and the other branches of Natural Philosophy teach us that inanimate bodies are either United with or acted upon by principles which are not material.

By Matter I understand what is meant by that Term by all Natural Philosophers, to wit, that ⟨it⟩ is a Substance impenetrable to other Matter divisible into parts, capable of being moved by any force impressed, with a Motion proportional to that Force, but incapable in itself of c⟨h⟩anging its state either from rest to motion or from motion to rest. This is the Notion of Matter on which the whole Fabrick of Natural Philosophy is built. These qualities of Matter have been proved by the most compleat Induction of Experiments and Observations. And the Science of Natural Philosophy is built upon them as Axioms. And if they be false the whole Fabrick of that Science falls to the Ground, & must be raised from a New & more stable Foundation. Dr Priestley indeed perceiving that this Notion of Matter did not well agree with his favourite doctrine of Materialism, has adopted another | that Matter is neither impenetrable nor Inert, & that it has inherent powers of Attraction and Repulsion which are so essential to it that without them it cannot exist. This new Notion of Matter I endeavoured to refute in the Discourses I already aluded to, which I am not now to resume. I endeavoured likewis to shew that this Notion of Matter if true, would overturn the whole of Sir Isaac Newtons Philosophy. This Dr Priestley seems not to have been aware of, because he pretends to build his Notion of Matter upon Sir Isaac Newton's rules of philosophizing

I shall therefore now (that I may not repeat what I formerly delivered) take it for granted that Matter is what all natural Philosophy proves it to be an inert impenetrable divisible & moveable Substance. And I say that from this we may conclude that even the inanimate bodies we perceive are either united to or acted upon by active substances that are immaterial.

All the Matter that falls within our Notice is acted upon by Gravitation. This is an active principle which cannot be inherent in inert Matter. It is therefore a force impressed upon Matter, and is always called by Sir Isaac Newton who discovered it, an *impressed Force* It is evident that an inert Body once put in Motion would always continue to move in the same streight line. When therefore it deviates from this direction every Moment and moves sometimes faster & sometimes slower as the Planets & Comets do, being passive in all these deviations it must be acted upon by a force constantly impressed upon it. A force constantly impressed upon matter supposes some Agent that has power to produce Motion in Matter. For Action can no more be without an ⟨Agent⟩ than qualities without a Subject. All the Attempts that have been made to account for Gravitation by the impulse of Subtile Matter are so weak as not to deserve a refutation. We must therefore conclude that the Gravitation of Matter is effected either immediately or ultimately by the power of some Being that is not Material.

What has been said of Gravitation may with equal Reason be applied to many other forces by which the parts of Matter are acted upon. Such as the force by which the parts of hard bodies cohere, corpuscular Attraction in the parts of Liquid bodies. Elasticity. both in hard bodies & in elastick Fluids, Magnetism Electricity, and all those affinities & hostilities which Chemistry presents to our View. These are all impressed Forces, and must immediately or ultimately ⟨be⟩ impressed by something that is not Matter.

In all these instances however of the powers that act upon inanimate bodies there is no Evidence of any such Relation between the body acted upon and that which acts upon it, as that they can be said to be united in one compound. In Gravitation for instance though we commonly consider the greater body as that ⟨which⟩ attracts and the less as that which is attracted, yet it is evident that both are attracted with equal force and that which produces the attraction is equally related to both and cannot be a part in the composition of either. It is so in all the Forces impressed upon inanimate bodies. They have all an equal Relation to different Bodies, which are equally acted upon. Hence when in popular language we attribute such action to the bodies themselves ⟨it⟩ is an universal law that there is a reaction equal and contrary to every action of Bodies on each other. When this Action and Reaction is conceived as it really is, an impressed Force it is one Action equally related to both.

Thus I think it appears that inanimate Matter is acted upon by some Agent or Agents that are not Material. And I think we have equal Reason to conceive that Animated Matter in Vegetables and in animals of every degree, is united to some Substance that is immaterial. In every Animal and even in every vegetable there is something which we call Life & something we call Death, and the Effects of Life & of Death in both are so similar as to give ground to conceive a similarity in the Causes of these Effects. In every living animal and vegetable there is a very curious Organization of fluid and Solid parts containing & contained by which it is fitted for its vegetable or animal Functions. When this organization began & how it is formed are things beyond the comprehension of our faculties, but it is evident that it could not be formed by inert Matter. When it becomes perceptible to us, we see at the same time an intestine Motion, which must | have a cause that is not material. It has the power to draw Nourishment from dead Matter so as to make it a part of itself, to expand in its bulk till it comes to its full stature, to produce after its kind, to extend its roots in quest of Nourishment and to bend its head in order to receive the rays of Light which are necessary to its verdure and health. One effect of Life even in Vegetables is that that organization which is necessary to the performance of the vegetable functions is preserved. But no sooner is Life at an end than that heat and moisture which was salutary in Life, becomes the mean of dissoluting of the whole Organization. It is now well known that animal and even vegetable Bodies when Life is gone are prone to be resolved into their principles & to have all organization destroyed. by various ways, by putrefaction, by fermentation, by digestion in the stomach of other Animals but that Life preserves the Organization from all these means of its destruction.

XIa
Philosophy of Substances & Attributes

2131/2/
III/13,
C verso

5   What ancient Philosophers called Matter & Form
Substances not immediately perceptible by the external Senses
Union of Substances very different
The Manner & Effects of such Unions above our Comprehension.
The proper Notion of Matter. taken for granted in this Discourse
10 Matter may have qualities which we know not but can have none which are inconsistant with those we know
All the bodies we perceive are acted upon or united to some thing that is not Matter.

Notes

EXPLANATORY NOTES
Natural History

MS 2131/3/14

Reid's notes on volume II of Buffon's *Histoire naturelle* are taken from chapters 1 ('Comparaison des animaux, des végétaux & des minéraux'), 2 ('De la reproduction en général'), 4 ('De la génération des animaux') and 5 ('Exposition des systèmes sur la génération').

83/21–2   Buffon attacked Linnaeus' *System naturae* (Leiden, 1735).

83/30   The Latin tag is taken from Horace's *Satires*, I. v. 100, and can be translated as 'Tell that to the Marines!'

84/36–85/3   Buffon discusses: Louis Bourguet, *Lettres philosophiques sur la formation des sels et des crystaux et sur la génération et le mechanisme organique des plantes et des animaux* (Amsterdam, 1729); Thomas Burnet, *Telluris theoria sacra* (London, 1681); G. W. Leibniz, 'Protogaea autore G.G.L.', *Acta eruditorum* (Leipzig, January 1693), 40–2; John Ray, *Miscellaneous discourses concerning the dissolution and changes of the world* (London, 1692) (which was published in revised form as *Three physico-theological discourses*, (London, 1693)); Nicolaus Steno, *De solido intra solidum naturaliter contento dissertationis prodromus* (Florence, 1669); William Whiston, *A new theory of the earth, from its original to the consummation of all things* (London, 1696); John Woodward, *An essay toward a natural history of the earth* (London, 1695); and a Latin dissertation on the origin of mountains and the theory of the earth read by Johann Jacob Scheutzer before the Académie des Sciences in 1708 reported in the *Histoire de l'Académie Royale des sciences. Année MDCCVIII. Avec les mémoires de mathématique & de physique pour la même année* (Paris, 1709), pp. 30–3.

85/6   René Antoine Ferchault de Réaumur, 'Sur les coquilles fossiles de quelques contons de la Touraine, & sur les utilités qu'on en tire', in *Histoire de l'Académie Royale des sciences. Année*

M.DCCXX. *Avec les mémoires de mathématique & de physique pour la même année* (Paris, 1722), pp. 400–16.

86/27 Reid has clearly written 'intrasusception', but the correct form is 'intussusception', which the *Oxford English dictionary* defines as the 'taking in of foreign matter by a living organism and its conversion into organic tissue'.

87/13–16 Although Buffon provides only two specific references, the relevant texts include: William Harvey, *Exercitationes de generatione animalium* (London, 1651); Marcello Malpighi, *Dissertatio epistolica de formatione pulli in ovo* (London, 1673); Antonio Vallisneri, *Opere fisico-mediche stampate e manoscritti del kavalier Antonio Vallisneri*, 3 vols (Venice, 1733); Regnier de Graaf, *De mulierum organis generationi inservientibus tractatus novus* (Leiden, 1672); Anthony van Leeuwenhoek, *Opera omnia*, 4 vols (Leiden, 1722).

MSS 2131/6/V/12 and 6/V/35

These notes are based on chapters 8 ('Réflexions sur les expériences précédentes'), 9 ('Variétés dans la génération des animaux'), and on the sections 'Des sens en général' and 'De l'age viril. Description de l'homme' of the 'Histoire naturelle de l'homme' in volumes II and III of Buffon's *Histoire naturelle*.

88/2 The *Oxford English dictionary* defines 'cicatrice' as the 'scar of a healed wound' or 'a scar-like mark or impression'.

88/13–14 John Turberville Needham, *An account of some new microscopical discoveries* (London, 1745); id., *Nouvelles déscouvertes faites avec les microscope* (Leiden, 1747).

88/22 Pierre-Louis Moreau de Maupertuis, 'Observations et experiences sur une des especes de salamandre', in *Histoire de l'Académie Royale des sciences. Année M.DCCXXVII. Avec les mémoires de mathématique & de physique pour la même année* (Paris, 1729), pp. 27–32.

91/9 James Parsons, *Human physiognomy explain'd: in the Crounian lectures on muscular motion. For the year MDCCXLVI. Being a supplement to the Philosophical Transactions for that year* (London, 1747).

MS 2131/3/II/16

Reid's notes indicate that he read no further than Part XI, chapter 6 of Bonnet's *Contemplation*.

Explanatory Notes 247

93/7-8 Charles Bonnet, *Considerations sur les corps organisés* . . ., 2 vols in 1 (Amsterdam, 1762); id. *Essai analytique sur les facultés de l'âme* (Copenhagen, 1760).

97/24 Bonnet's work originally appeared as *La palingénésie philosophique, ou idées sur l'état passé et sur l'état futur des êtres vivans*, 2 vols (Geneva, 1770).

97/30 Charles Bonnet, *Traité d'insectologie; ou observations sur les pucerons* (Paris, 1745).

97/32-3 Charles Bonnet, *Essai de psychologie; ou, considerations sur les operations de l'ame, sur l'habitude et sur l'education. Auxquelles on a ajouté des principes philosophiques sur la cause premiere et sur son effet* (London, 1745).

97/34-6 Lazzaro Spallanzani, *Prodromo di un opera da imprimersi sopra le riproduzioni animali* (Modena, 1768).

98/12-13 Jacques Christophe Valmont de Bomare, *Dictionnaire raisonné universal d'histoire naturelle*, new edn, 9 vols (Paris, 1775).

98/26-9 François-David Hérissant, 'Eclaircissemens sur l'ossification', *Histoire de l'Académie Royale des sciences. Année M.DCCLVIII. Avec les mémoires de mathématique & de physique, pour la même année* (Paris, 1763), pp. 322-36; see also the report 'Sur l'ossification' on pp. 31-6 for further discussion of Hérissant's memoir.

98/30-1 Charles Bonnet, *Recherches sur l'usage des feuilles dans les plantes* (Gottingen and Leiden, 1754).

Physiology

MS 2131/7/II/2

102/18-22 Experiments such as these featured in the dispute between Albrecht von Haller and Robert Whytt over the concepts of sensibility and irritability. The two key works in this dispute were Haller's 'De partibus corporis humani sensilibus et irritabilibus', *Commentarii Societatis Regiae scientiarum Gottingensis*, 2 (1752), 114-58 (translated as *A dissertation on the sensible and irritable parts of animals* (London, 1755)), and Whytt's 'Observations on the sensibility and irritability of the parts of men and other animals', in his *Physiological essays* (Edinburgh, 1755); see also p. 64, n. 78 above.

MS 3061/2

105/4 An allusion to Galen's *De usu partium corporis humani* ('Of the uses of the parts of the human body'); compare *Intellectual powers*, Essay VI, ch. vi, p. 630.

108/2 J. T. Desaguliers discusses experimental demonstrations of this kind in his *A Course of experimental philosophy*, 2 vols (London, 1734–44), I, 255–9. Significantly, Desaguliers mentions that he went to observe similar feats of strength performed by a German strongman in about 1719 with Sir John Pringle, Lord Tullibardine and the physician Alexander Stuart. Reid was acquainted with Stuart through his close friend, the Aberdeen mathematician John Stewart.

109/16-18 Compare *Inquiry*, ch. 6, sec. 10.

116/32 Reid refers to James Jurin, 'An essay upon distinct and indistinct vision', in Robert Smith, *A compleat system of opticks in four books, viz. a popular, a mathematical, a mechanical, and a philosophical treatise. To which are added remarks upon the whole*, 2 vols (Cambridge, 1738), II, 115–71 (esp. pp. 136–44).

Materialism

MS 2131/3/I/25

Reid's notes are taken from Priestley, *Institutes*. The notes from Hartley's *Observations* derive from the preface and propositions 1, 2, 6–10, 12, 14, 19, 21–2, 80, 86–9.

127/12-14 Priestley, *Institutes*, I, 56–7.
127/15-17 Priestley, *Institutes*, I, 15–16.
127/26-32 Priestley, *Institutes*, I, 67–8.
128/20-2 In the preface to the *Observations*, David Hartley traces the genesis of the work to his reading of John Gay's 'A dissertation concerning the fundamental principle and immediate criterion of virtue. As also, the obligation, and approbation of it, with some account

of the origin of the passions and affections', which was prefixed to William King, *Essay on the origin of evil*, trans. Edmund Law (London, 1731); David Hartley, *Observations on man, his frame, his duty, and his expectations*, 2 vols (London, 1749), I, v.

129/10–14 Reid registers Hartley's statement that we cannot understand the causal connection between vibrations in the brain and our sensations and ideas; Hartley, *Observations*, I, 33–4.

130/32 Reid repeats Hartley's mistake in referring to the 'Rule of False', when he should have said the 'Rule of False Position'; compare Hartley, *Observations*, I, 345.

131/8–18 Hartley, *Observations*, I, 350–1.

131/14–15 A reference to Aristotle's *Categories*, and John Wilkins, *An essay towards a real character and a philosophical language* (London, 1668).

131/26–132/13 Hartley, *Observations*, I, 359.

132/20–2 Berkeley attacked Locke's account of abstract ideas in the introduction to *A treatise concerning the principles of human knowledge* in *The works of George Berkeley Bishop of Cloyne*, ed. by A. A. Luce and T. E. Jessop, 9 vols (Edinburgh, 1948–57), II, 25–40. Significantly, in one of the earliest of Reid's extant manuscripts (dating from 1738), he defended Locke's account against the criticisms of Berkeley and Peter Browne; see AUL MS 3061/10, 2r–v.

MS 3061/9

As discussed above, pp. 35–6, Reid wrote the 'Miscelaneous Reflections' in response to Priestley, *Hartley*. Reid also makes reference in this manuscript to Priestley, *Examination*, and to the third volume of Priestley, *Institutes*, which appeared in 1774.

An edited and abbreviated version of the 'Miscelaneous Reflections' appeared in the *Monthly review*, 53 (1775), 380–90 and 54 (1776), 41–7. The identification of the authorship of this review is made by I. C. Douglas, in 'David Hartley and the perfectibility of man' (unpublished M.Phil. thesis, University of Leeds, 1989), p. 43. I would like to thank Ian Douglas for bringing the published version to my attention, and for supplying me with a copy of his thesis.

133/2 Reid refers to the exchanges between Samuel Clarke and Anthony Collins; see above, p. 75, n. 180.

133/18 Reid probably alludes to Book I, chs 4–8 of Aristotle's *Posterior analytics*, 73a–75b.

136/18 Priestley, *Hartley*, pp. xxii–iii. Locke discussed the association of ideas in *An essay concerning human understanding*, ed. P. H. Nidditch (Oxford, 1975), II. xxxiii; this chapter was added to the fourth edition of the *Essay*. Priestley also notes John Gay's 'A dissertation concerning the fundamental principle and immediate criterion of virtue', as cited above in the note to 128/20-2.

136/23-30 Reid refers to Aristotle's *On Memory and Reminiscence*, 451a–453b; Hobbes's discussion of the causes of the 'coherence' of our ideas in the fourth chapter of *Human nature, or the fundamental elements of policy* (see *The English works of Thomas Hobbes of Malmesbury*, edited by Sir William Molesworth, 11 vols (London, 1838–45), IV, 14–19); and David Hume, *A treatise of human nature*, I. i. 4 ff.

138/11 Hume, *Treatise*, I. i. 3.

138/21-3 For Hartley, a 'vibratiuncle' was a weaker form of vibration in the medullary substance of the brain which accompanies an idea; Hartley, *Observations*, I, 58–9, and above, 129/20-2.

139/25 Locke, *Essay*, IV. i. 1–7.

141/13 Locke, *Essay*, II. xxi. 30.

141/26 Priestley had attacked Reid's notion of instinctive principles of the mind in the *Examination*, pp. 9–24, 74–95.

142/38 Compare chs 1–3 and 6 of Hobbes's *Human nature* (*English Works*, IV, 1–14, 26–30); compare also chs 1–2 of *Leviathan* (*English Works*, III, 1–10).

144/34 Quintus Roscius (d. *c*. 62 BC) was the famous Roman comic actor, while David Garrick (1717–79) dominated the English stage in the middle decades of the eighteenth century.

145/5-10 Priestley, *Hartley*, pp. xxxv–vi.

146/10-23 Priestley, *Institutes*, III, xxi–ii.

146/35-9 Hartley, *Observations*, I, vi.

147/3-8 Hartley, *Observations*, I, 6.

147/19-27 Priestley, *Hartley*, pp. iii, v.

150/23-6 Hartley, *Observations*, I, vi.

151/10-15 Bishop Bramhall drew particular attention to Hobbes's scruples about the publication of his *A treatise of liberty and necessity*; see John Bramhall, *A defence of true liberty from antecedent and extrinsicall necessity, being an answer to a late book of Mr. Thomas Hobbs of Malmsbury, intituled, a Treatise of liberty and necessity* (London, 1655), 'To the Reader', and pp. 92, 118, 247, for both Hobbes's own texts and Bramhall's comments.

152/36-8	Hartley, *Observations*, I, 7.
152/39–153/2	Hartley, *Observations*, I, ii.
153/8-11	Hartley, *Observations*, I, 8.
153/13	Hume, *Treatise*, I. i. 1.
154/9	Priestley, *Hartley*, p. xix.
154/17-21	Reid quotes the General Scholium to Newton's *Principia*: 'I frame no hypotheses; for whatever is not deduced from the phenomena is to be called an hypothesis; and hypotheses, whether metaphysical or physical, whether of occult qualities or mechanical, have no place in experimental philosophy' (Newton, *Principia*, II, 547).
154/27	Reid means the rule of false position; compare 130/32.

MS 2131/3/III/23

This manuscript appears to be a preliminary draft of sections of Essay II, ch. 3, 'Hypotheses concerning the nerves and brain', of *Intellectual powers*.

158/21	Locke, *Essay*, II. iii. 1.
158/29–159/11	Compare *Inquiry*, ch. 7 and Essay II, chs 4, 7–8 of *Intellectual powers*. Reid probably refers to: Aristotle, *On the soul*, 418a–424b; Diogenes Laertius, *Lives of Eminent Philosophers*, Bk. x (on Epicurus); and Plato, *Theaetetus* and *Republic*, Bk. VI.
158/34-9	For a clearer formulation of the point Reid is trying to make see *Intellectual powers*, Essay I, ch. 1, p. 25.
159/36	Reid alludes to Part One, sec. 32 of Descartes' *The passions of the soul*; see Descartes, I, 340.
160/30-1	'for imagining is simply contemplating the shape or image of a corporeal thing' (Descartes, II, 19).
160/32-3	'Now all that the intellect does is to enable me to perceive the ideas which are subjects for possible judgements' (Descartes, II, 39).
160/34-6	'In order to be free, there is no need for me to be inclined both ways; on the contrary, the more I incline in one direction . . . because I clearly understand that reasons of truth and goodness point that way . . . the freer is my choice' (Descartes, II, 40).
160/37-9	This passage is from the Sixth Meditation: 'Hence the fact that I can clearly and distinctly understand one thing apart from another is enough to make me certain that the two things are distinct, since they are capable of being separated, at least by God' (Descartes, II, 54).

MS 3061/12

Reid's notes are taken from Priestley, *Disquisitions*.

MS 3061/13

Reid's notes are taken from Priestley, *A free discussion*; Priestley *Disquisitions*; Priestley, *Hartley*; and the third volume of Priestley, *Institutes*.

164/3   Joseph Berington, *Letters on materialism and Hartley's theory of the human mind, addressed to Dr. Priestley* (London, 1776). Berington (1746–1827) was a Catholic priest, and a vocal opponent of Priestley.

MS 3061/23

164/18-20 Priestley, *A free discussion*, pp. 241–2.
166/1-2 Compare Locke, *Essay*, II. xiii. 19 and II. xxiii. 2.
166/11-12 Priestley, *Institutes*, III, 12, and *Examination*, p. 155; compare 163/3–7 and 163/29–30.
166/12-14 Priestley, *Examination*, p. 321; compare 163/9–11.
166/14-17 Priestley, *Institutes*, III, 12–13, and *Examination*, pp. lviii–lix.
166/36-9 Priestley, *Examination*, p. 5.
167/8-10 Priestley, *Disquisitions*, p. 2.

MS 3061/14

168/28-32 Priestley, *Disquisitions*, p. 2.
168/38-9 Reid's original formulation echoes *Intellectual powers*, Essay II, ch. 3, p. 87; the literary allusion is to Alexander Pope's *The rape of the lock* (1714).
169/16-17 Newton introduced the phrase 'as far as possible' in the third edition of the *Principia*; see above p. 72, n. 153.
170/15-19 Inaccurately quoted. The sentence reads: 'It will perhaps be said, that the particles of which any solid atom consists, may be conceived to be placed close together, without any mutual attraction between them' (Priestley, *Disquisitions*, p. 6).

MS 2131/3/I/17

These notes are taken from Benedetto Stay's (1714–1801) *Philosophiae recentioris a Benedicto Stay . . . versibus traditae libri X . . .*

Explanatory Notes

cum adnotationibus, et supplementis P. Rogerii Josephi Boscovich S.J., 2 vols (Rome, 1755–60). The first edition of this work appeared in Venice in 1744, and a second was published in Rome in 1747.

172/4–5 'Indeed, it is my opinion that body is constituted entirely by points unextended and separate from each other.'

172/7–8 The Latin text of Newton's first rule reads: 'Causas rerum naturalium non plures admitti debere, quam quae et vera sunt et earum Phenomenis explicandis sufficiunt.'

172/9–10 The passage referred to reads: 'On my view of matter, as composed of indivisible, unextended and individually distinct points (we shall see this, as I have indicated, in Book 10, and also that empty space is infinitely greater than matter), a difficulty of that sort has far less force, and those seeming co-penetrations can be had much more easily without true co-penetration.'

MS 3061/1/4

James McCosh was the first to draw attention to 'Some Observations On the Modern System of Materialism' and the manuscripts related to it (presently catalogued together in the 3061 collection). These manuscripts were lent to McCosh by Francis Edmond in 1864; see AUL MS 2814/1/80, and McCosh, p. 473.

173/12–13 It should be noted that Reid does not refer to the second, enlarged edition of the *Disquistions* published in 1782; on the significance of this point see above pp. 38–9.

178/7–10 'I frame no hypotheses; for whatever is not deduced from the phenomena is to be called an hypothesis; and hypotheses, whether metaphysical or physical, whether of occult qualities or mechanical, have no place in experimental philosophy' (Newton, *Principia*, II, 574).

178/12–15 Reid quotes the description of the practice of experimental philosophers in Cotes's preface to the second edition of Newton's *Principia*: '. . . they assume nothing as a principle, that is not proved by phenomena. They frame no hypotheses, nor receive them into philosophy otherwise than as questions whose truth may be disputed' (Newton, *Principia*, I, xx).

178/17–18 Priestley, *Examination*, p. lix.

181/8–10 Locke, *Essay*, I. iv. 18.

182/16-29 Priestley, *Disquisitions*, pp. 1–2.

186/15-18 Priestley, *Disquisitions*, p. 2.

186/25-7 'We are to admit no more causes of natural things than such as are both true and sufficient to explain their appearances' (Newton, *Principia*, II, 398).

189/1-2 Reid uses the reading of the rule contained in the first two editions of the *Principia*. In the third edition (which he does not cite), the Second Rule read 'Ideoque effectum naturalium ejusdem generis eaedem assignandae sunt causae, quatenus fieri potest'; see above p. 72, n. 153.

191/15-18 'The qualities of bodies, which admit neither intensification nor remission of degrees, and which are found to belong to all bodies within the reach of our experiments, are to be esteemed the universal qualities of all bodies whatsoever' (Newton, *Principia*, II, 398).

192/16-28 Priestley, *Disquisitions*, p. xxxviii.

193/21-5 Priestley, *Disquisitions*, pp. 3, 4.

196/20-2 Priestley, *Disquisitions*, p. 13.

197/19-22 Priestley, *Disquisitions*, pp. 6–7.

199/29-32 Priestley, *Disquisitions*, pp. 6–7.

199/37–200/2 A paraphrase rather than a quote from sec. 1 of the *Disquisitions*.

203/23-7 Priestley, *Disquisitions*, p. xxxviii.

204/2-6 Priestley, *Disquisitions*, pp. xxxvii, xxxviii.

204/20 The 'Illustrations' form part of Priestley, *A free discussion*.

207/1-3 Priestley, *Disquisitions*, p. 2.

209/11-12 Reid quotes the General Scholium to the *Principia*: 'But hitherto I have not been able to discover the cause of those properties of gravity from phenomena, and I frame no hypotheses' (Newton, *Principia*, II, 574).

210/38–211/1 Reid quotes Newton's third letter to Richard Bentley (*Isaac Newton's papers and letters on natural philosophy*, ed. I. B. Cohen (Cambridge, 1958), 302–3).

211/27-36 Priestley, *A free discussion*, pp. 230–1.

212/7-12 Priestley, *A free discussion*, p. 231.

212/33-6 Priestley, *A free discussion*, p. 231.

214/5-8 Reid quotes Newton's 'Advertisement' to the 1717 Latin edition of the *Opticks*, which in the English version reads: 'And to shew that I do not take Gravity for an essential Property of Bodies, I have added one Question concerning its Cause, chusing to propose

Explanatory Notes

it by way of a Question, because I am not yet satisfied about it for want of Experiments' (Newton, *Opticks*, p. cxxiii).

214/9–15 Priestley, *A free discussion*, pp. 231–2.

214/26–8 Reid quotes the General Scholium to the *Principia*: 'This most beautiful system of the sun, planets, and comets, could only proceed from the counsel and dominion of an intelligent and powerful Being' (Newton, *Principia*, II, 544).

MS 2131/2/III/13

230/32–231/33 For the references in these paragraphs see above, p. 75, nn. 179–80.

TEXTUAL NOTES
Natural History

MS 2131/3/II/14

83/29–30	Hence . . . Apella *added in left margin*
85/14–15	of Dry land] of Earth
85/21	room for *added*
86/7–8	which nourish] of
86/9	after *added*
87/9	perhaps *added*

MSS 2131/6/V/12 and 6/V/35

Although the two parts of this text are currently catalogued separately, they do in fact belong together. In the bottom right-hand margin of MS 2131/6/V/12,2v, Reid has marked a 'B', and a corresponding 'B' is to be found marked in the upper-left hand margin of MS 2131/6/V/35,1r. The text is continuous.

87/21	Men] Men Dogs Cows (*deleted word illegible*) Rabbits & other Males there were found
88/26	the Pidgeon *added*
88/34–5	Bees . . . neutral *added*
89/20–1	A . . . Cock *added in right margin*
90/38	& entangled *added*
91/21	past *added*
92/3	reflects . . . & *added*
92/18	cooperating] either cooperating

MS 2131/3/II/16

For the dating of this manuscript see above p. 16. The notes on 2r–3v from Bonnet's La Palingénésie philosophique *appear to have been taken at a different time because the ink is darker and the nib of Reid's pen is coarser than in the previous section.*

258 Notes

93/3   Reid has used a sheet on which the following lines had been written and cancelled:

Of the Subjects of Active Power

Having endeavoured to shew that Active Power is not a word without any Meaning, & that although we cannot give a Logical Definition of it, although our Notion of it be onely (onely *added*) relative, although it be not an Object either of Sense or of Consciousness, yet Mankind have and allways had such a Notion of Active Power as that they speak and Reason about it with understanding. We are now to consider the Subjects of Active Power, that is to what Being we have reason to ascribe Active Power.

93/21   all *added*
96/19–20   Polype d'Entonnoir *added*
97/19   Reid has added a those *to* pious *which I have omitted in order to preserve the syntax of the sentence*
97/30   a Paris *added in right margin*

Physiology

MS 2131/7/II/19

101/3   of Physiologists *added*

MS 2131/7/II/2

101/29   Medicines] things
102/16   constant *added*
103/1   last *added*

MS 3061/2

103/13   having] have
103/14   two *added*
103/28–9   Some . . . first] First some general Observations
105/2   in *added*

Textual Notes

105/28 makes the whole] Muscles makes the whole length
105/35 even when this is] though the last be
105/37 or are bent (over them *deleted*) in order to change their direction *added*
106/2   to change their direction *added*
106/6   through *added*
106/16 the . . . length] its contraction
106/28 any of the contiguous parts. *added*
106/37 Body] human Form
106/39 Contour or *added*
107/1   The beauty of its Form] Its beauty
107/24 acts] is
107/25 a *added in* a case
108/4   Roller] Cannon
108/7   indeed *added*
108/9   of . . . made] what soft & pliable materials they are made of
108/18 be exerted] be contracted
108/20–1 skill . . . Artfully] ability to manage this machinery so skillfully
108/25 that he *added*
108/30 with . . . Printer] perfectly
108/31 certainly *added*
109/1   most *added*
109/5   independent . . . another *added*
109/20–1 the . . . is not.] not the Cause but what is
109/21 Work] Work done with so much contrivance and Design
109/27 the interposition *added*
109/35 By no means. *added*
110/5   sudden *added*
111/1   an uneasy] a
111/15–17 This . . . the Ballance. *added in right margin of p. 10*
111/39 contracted,] contracted, & no others
112/24–6 *The order of the two italicized sentences is reversed in the manuscript. Reid indicates that he wanted them to read as they appear here.*
112/27 there are] the body has
112/27 Spine] back
112/30 like . . . links *added*
112/37 in any position *added*
112/39 That *added*

113/6	keep] keep such
113/11	to act in] for
113/12	in number *added*
113/13	intend] will
113/14	at least *added*
113/15	muscles *added*
113/19	that which we intend] the direction intended
113/21	accurately adjusted] accurate
114/7	in some persons *added*
114/9	through] of
115/2	both *added*
115/18–25	This Consent . . . the Consequence. *added in left margin of p. 15*
115/32	a young] the
115/35	half] one half
116/20	like a Gloworm *added*
116/30–1	by . . . Axis *added*
116/35–6	& that . . . Muscle *added*
117/4	Combination] Mechanism
117/6	shall *added*
118/4	if *added*
118/6	here . . . Stand. *added*
119/20–1	and . . . Intestines *added*
119/22	its *added*
120/18	different] different of which that
120/30	onely *added*
122/1	by *added*
122/27	make him *added*
123/7	some] the
123/22	at one . . . another] more or less
123/28	to have *added*
124/11	on . . . Motions *added*
124/13	Age] Old Age
124/25	within *added*

Materialism

MS 2131/3/I/25

127/8	page 8. *added*
128/21	the Revd *added*
128/34–5	hurt . . . another] injure them
129/7	*After* body. *Reid has deleted the following passage*: A Sound State of the Brain is no less necessary to the vital & involuntary motions of the body than to the Operations of the Mind,
131/18	other *added*

MS 3061/9

In this manuscript Reid put a number of his page references in the margins and, in the interests of readability, I have recorded these as footnotes.

132/35	this . . . great] it of great and general
133/30	Discovery] one
134/7–8	and . . . two] but an elementary Body and a Mind
134/12	consession *added*
134/13	has been one of the chief] is one of the main
134/18	faith] Arguments
134/22–3	we apprehend *added*
135/11	Limbus] Dungeon
134/14	says he *added*
135/21–3	Whereas . . . them. *Reid had great difficulty in formulating this sentence to his satisfaction, and there are a number of stops and starts.*
135/27–37	Secondly . . . told. *added in left margin*
136/12	the faculties of *added*
136/37	as little] not
137/20	that *added*
137/27	it cannot be doubted] there can be no doubt
137/29	either *added*

262   *Notes*

138/18–19 held to be *added*
139/31   perfect *added*
141/9–10 or to] or
141/16   nor . . . Desire *added*
141/18   without . . . granted] takes it for granted, without offering any proof
141/33   however *added*
144/5   which] by which
144/25   but two things here deserve to be noticed. *added*
145/11–12 assemblage] combinations
146/27–9 "thrown . . . 2d) *added in right margin*
148/9   Pattern] Example
149/6   farther] more
150/6   may be found] is
150/18–19 both . . . and *added*
150/21   them] his conjectures
150/25   had begun] began
150/28   even *added*
150/33   their Doctrine] the Doctrine
151/2–3   to believe the contrary *added*
151/35   left *added*
152/14   from] of
152/21   either . . . with] with
152/38   & Motion *added*
153/35   occasions] attends
154/9   of a Mind *added*

MS 2131/3/III/23

This manuscript is written on the unused portions of a printed letter dated 18 April 1778 from Matthew Baillie, son of the recently deceased Professor of Divinity at Glasgow, James Baillie.

154/37   brought to explain] they are assigned as the causes
154/38   *After* 'perception' *Reid has deleted the following passage*: It is to be observed that Dr Hartly as he seems sceptical with regard to the souls immateriality, desires not to be interpreted (understood *uncancelled*) so as not to oppose the immateriality of the Soul altho' he thinks his theory overturns all the Arguments which are usually brought for that doctrine and that it follows from his Theory

that if matter could be endowed with the most simple kinds of Sensation, it might also arrive at all that Intelligence of which the human Mind is possessed

155/11   even *added*
155/16   most *added*
155/32-3 when brought *added*
155/36   For . . . Sensations *added*
155/39   in degre onely] onely in degre
156/2   innumerable] a variety of
156/8   How shall we find] There must be
156/11   and *added*
156/13   primarily affect one part of the brain or another] may enter the Brain at one part or another (*uncancelled*)
156/14-15 according . . . Nerves *added*
156/18   four Variations of Vibrations] Variations
156/18   should it not] It should
156/20   *After* kind *Reid has deleted*: This by no means tallies with the Phenomena.
156/23-4 & . . . than] each of them having Sensations of variety
156/27   *Reid has written* Senstations
156/28   *Reid has written* not onely (*illegible deletion*) but But *above the words* very hard . . . by heaping supposition
156/29   not onely] and
156/35   Sound] hearing
156/36   First *added*
157/5   or discord *added*
157/7-8 it . . . Men *added*
157/11-12 If . . . Substance] If any thing like this could be shewn in the Accounting for the other Senses by Vibrations of any Elastick Substance whatsoever (whatsoever *added and the whole of the original version of the passage left uncancelled*)
157/13   we are told] men pretend (*uncancelled*)
157/14-15 no man could ever shew] is neither known (*uncancelled*)
157/17   variety . . . the *added*
157/18   a System] an Account
157/23   that Connexion] their Conexion
157/24   this] the Existence of those Vibrations. We know the Existence of our Sensations by Consciousness but of the Existence of Vibrations in the Medullary Substance of the Nerves and Brain No proof has

264   *Notes*

157/25 yet been brought and therefore it is impossible to prove a Connexion between one and the other. As
157/25 Vibrations; & it] Vibrations; so it
157/33 therefore ... Hypothesis] that we can expect from this Hypothesis therefore
157/34 like] the same
158/1–2 find even] even find
158/27–8 have ... retained] have been with some Small Variations, to have retained
158/34 & his followers *added*
158/35 the objects of human Understanding] human knowledge (*uncancelled*)
159/3–4 because ... fluctuation *added*
159/13 as to] in
159/24 probable therefore that] probable that
159/30 Since] When
159/37–8 & ... are] all the other parts were
160/3 As] With regard
160/4 well established] evident (*uncancelled*)
160/6 whether] & whether
160/7 or ... have] &
160/9 left ... them] given over these Disputes
160/12 made ... prodigious] furnished with a great
160/14 & ... fitted] which it is fitted to produce
160/15 confine it to] place it in
160/17 & Effects *added*
160/19 from the brain *added* (*two added words after* brain *illegible*)
160/23–4 putrefaction ... bodies] Action of the Air and circumambient bodies which would soon reduce the whole to putrefaction this
160/24–5 anatomists and *added*

MS 3061/12

This manuscript also contains reading notes from Priestley's The doctrine of philosophical necessity illustrated *(1r–v, 3r–v), Bishop Butler's* Analogy of religion, natural and revealed *(2r), and an interesting note on Berkeley (4r).*

161/6 p 2 *added in right margin*

MS 3061/13

All of Reid's page references on 1r of this manuscript are written in the left margin and (following his practice on 1v–2r) I have incorporated them into the text.

162/26   *Reid has written* a 4 *in the left margin*
163/4–5 *Reid has written* b1 *in the right margin*
163/9   Dr Price *added. Reid has written* b2 *in the right margin*
163/17–18 *Reid has written* b3 *in the right margin*
163/24–5 *Reid has written* b4 *in the right margin*
163/29–30 *Reid has written* b5 *in the right margin*

MS 3061/23

164/10   said in] said times innumerable in
164/11   have] have very generally
164/12   generally *added*
164/17   ancient *added*
164/25   Principles] the Principles
164/28   Sum] Substance
164/38   is it . . . it] Inert, & in reality
164/39   inherent *added*
165/6   inherent *added*
165/16   Inventor] sole Inventor
165/28   to *added*
166/4   Dr *added*
166/18–19 & . . . Materialism *added*
166/21   therefore *added*
166/22   than he has given *added*
166/27   by just inference concluded] justly inferred
166/28–9 & . . . assents *added*
166/30–3 concluded . . . Sophism.] That there is no such thing as Matter in the Universe (*revised wording added in right margin*)
166/32–3 I . . . reality] in reality is
166/33–4 of . . . Materialism *added*
167/10   onely *added*
167/15   of them *added*
167/16–17 occult. . . Vortices of] System of Aristotle or
167/17   ⟨the System⟩] that (*deleted*)

MS 3061/14

168/36 in . . . than] Sir Isaac Newton thought as necessary as
168/37 requires] makes
168/38–9 & the Vibrations & Vibratiuncles of Dr Hartley] Silphs & Gnomes of Mr Pope
169/3 want] are Fictions & want
169/4–7 To . . . design. *added on 2v, with insertion point marked* A
169/14–15 Such . . . paraphrase. *added*
169/16–34 Sir . . . it. *added in margin over the fold of 1r–2v.*
169/16 apply] use (*uncancelled*)
169/17–18 as . . . applied (*insertion marked* X)
169/18 this would not] nor would this
169/25 & therefore to have the same Cause *added*
169/27 those Effects that] such Effects as
170/33–4 or in any casual concourse *added*

MS 2131/3/I/17

171/37 Raguisina *added*
172/11 good or bad *added*
172/12 is not] cannot be
173/1 matter] bodies

MS 3061/1/4

MS 3061/1/4 provides a text of Reid's 'Some Observations On the Modern System of Materialism' which does not require elaborate editorial intervention. The manuscript is a fair copy apparently written in another hand, but it includes corrections made by Reid himself. My transcription thus appears to represent Reid's final wording, and will therefore serve most scholarly purposes. However, specialist researchers will still need to consult other related manuscripts in the 3061 collection for detailed evidence concerning the evolution of Reid's wording. Unlike 3061/1/4, MSS 3061/1/1–3 are all in Reid's hand. MS 3061/1/1 is an incomplete and heavily worked-over version of 'Some Observations'. MS 3061/1/2 is complete, and incorporates numerous corrections and additions. Further additions to MS 3061/1/2 are catalogued as MS 3061/24. MS 3061/1/3 is another incomplete version of 'Some Observations', which includes various corrections. MS 3061/18 is a draft of what became the third section of chapter one of 'Some Observations'.

The physical characteristics of MS 3061/1/4 differ strikingly from MSS 3061/1/1–3, and other material in the Birkwood Collection. MS 3061/1/4 consists largely of individual sheets of paper (roughly 21.2 cm. × 26.5 cm. or 8¼ in. × 10½ in. in size) which have been crudely stitched together. MSS 3061/1/1–3 are all written on paper which is markedly smaller in size, like that used elsewhere in Reid's manuscripts.

175/4	were] are
189/24–5	not . . . but *added*
190/17	in time past in like circumstances] in like circumstances in time past
196/1	hard] small
197/19	each other] one another
215/19–21	I . . . Fruits. *added in right margin*

MS 2131/2/III/13

This manuscript is made up of two parts. The first, which I have called 'A', consists of a single sheet of paper written on both sides. The second, which I have called 'B', consists of two folded sheets which comprise the sequence B1r–B4r. Another small sheet of paper, which I have called 'C', has come to be catalogued with parts A and B, although its contents are related to both MS 2131/2/III/13 and MS 2131/2/I/15. C recto contains a more detailed treatment of topics dealt with on A recto, and I have printed this below as manuscript Xa.

217/7	examine] refute
217/9	if Matter be what] Matter such as
217/10	Philosophers] Philosophy
217/11	it *added*
217/12	thinks . . . it] hopes to remove this prejudice by shewing that Matter
217/21	tried] examined
217/23–4	the Reasons he has assigned] Reason
217/25	be] were
217/35	therefore takeing] I shall therefore take
217/36	as a Doctrine which needs] conceiving that it must

217/37–8	& takeing it for granted] and I shall take it for granted in this Discourse
217/38	passive] inactive
218/5	necessarily *added*
218/13	the body which is the Subject of Dispute *added*
218/17	inanimate] dead
218/17–21	nothing . . . Man.] no real Quality in the whole which is not in the parts. When taken to pieces it is still the same thing it was, nothing is lost And when the parts are put together again in the same order it is just what it was (*uncancelled*)
218/24	heal its wounds *added*
218/27	that wh⟨ich⟩ (*manuscript damaged*)
219/7	according . . . Logick *added*
219/26	of . . . Bodies *added*
219/29	active *added*
220/7	the impulses of *added*
220/38–9	let . . . Matter *added*
221/4	objected] said
221/11–12	which . . . System *added*
221/18	appointed . . . purpose *added*
221/26	suspension] cessation
221/33	both *added*
222/11	and all things *added*
222/22	⟨will⟩ (*manuscript illegible*)
222/25	beyond] problematical & beyond
222/28	him] him for that purpose
223/12	the . . . to *added*
223/15	inanimated] inert
223/22	in all degrees of heat we are able to produce *added*
223/23	various combinations] a combination
223/24	animated . . . produced] concrete Bodies are formed
223/25–6	are no doubt] may be, and very probably are
223/32	combustion *added*
223/37	Many] Many of the
224/3	Being] System (*uncancelled*)
224/4	active *added*
224/9	symptom] part nor symptom
224/11	led *added*
224/32	really *added*

224/34	a *added*
225/19	living *added*
225/24-5	when disjoyned from] without the aid of
225/27	onely] probably
225/33	what . . . dictates *added*
225/35	not to] to no dead
226/1-2	peculiar . . . Matter] of the Elements of Matter peculiar to that Species.
226/15	destroy . . . & *added*
226/16	turn] form (*uncancelled*)
226/27	were digested in it] it digested
227/11-13	*It is unclear why Reid has enclosed this passage in stars*
227/12	a progeny *added*
227/17-19	constant . . . prove] which we before observed between the inanimate kingdom and things immaterial
227/23	occasionally . . . requires *added*
227/31	Heart . . . are] Heart is
227/32	sense I apprehend] Manner (*uncancelled*)
227/37	an Effect] a Fact
228/8	are the consequences of] follow from
228/21	the Living Principle within it] something internal
228/24	a Ventricle of *added*
228/26	muscular Fibres *added*
228/36	Thus] From what has been said
228/39	mean to express *added*
230/18	nor . . . Consequence. *added*
230/21	justly *added*
230/27	while . . . Matter *added*
230/33	or Conjecture *added*
231/9	compounded] made up

MS 2131/2/III/13 C recto.

232/23	a] the
232/26	or] and the parts may be

MS 2131/2/I/15

233/7	that Author] Dr Priestly
233/11	Philosophers . . . downwards] Philosophy
233/22	from . . . them *added*

233/23-4	it . . . Ground *added*
233/35	Substance is neither an Object] They are neither Objects
234/12	of the same kind] one and the same (*uncancelled*)
234/13	Mind . . . Body] the one and the other
234/15	one . . . other] the Attributes of one and those of the other
234/16	& in] &
234/26–235/7	*After* contrary *Reid has written* add A. *This seems to refer to the passage marked 'A' in 2131/2/III/13, C verso, and I have placed it here. The passage runs from* Nor *to* immaterial
234/36	the lowest degree *added*
234/36	Being] Beings
235/19	are . . . Nature *added*
236/26-7	of the animals existence *added*
236/38	all the Phaenomena of Life be the result of] it be onely
236/39	or . . . it *added*
237/36	Substances] things
238/1-2	which . . . affinity *added*
238/5	qualities] appearances
239/14	as . . . do *added*
240/3	which] who
240/4	⟨it⟩ *Reid has written* in
240/24	draw . . . Matter] assimilate dead matter (to assimilate *uncancelled*)
240/32	dissoluting] putrefaction and dissolution

INDEX

Aberdeen
  Gordon's Mill Farming Club, 56
  King's College, 3, 56, 74
  Marischal College, 58
  Philosophical Club, 3, 10, 20, 56
  Philosophical Society, x, 11, 40, 57
analogy of nature, 50, 230
Anderson, John, 60, 71, 76
animalcules, 87–9, 90
animalculism, 10
animism, 25, 27
anti-clericalism, 32
Aristotle, 83, 131, 133, 136, 137, 158, 159, 167, 187, 249, 250, 251
atheism, 7, 8, 32, 45, 47, 74, 132

Bacon, Sir Francis, 42, 43, 151, 154, 184, 187, 188, 191, 201
Baker, Henry, 11
Baxter, Andrew, 20
Beattie, James, 32, 35, 53, 55, 69, 74, 78
Bentley, Richard, 46, 74
Berington, Joseph, 252
Berkeley, George, 132, 163, 166, 174, 175, 176–80, 181, 182, 249
Boerhaave, Herman, 27, 66
Bolingbroke, Lord, *see* St. John, Henry
Bomare, Jacques Christophe Valmont de, 98, 247
Bonnet, Charles, 10, 16, 18, 27, 63, 93–8, 246–7
  Contemplation de la nature, 16, 17, 93–7, 246
  Palingénésie philosophique, 16, 63, 97–8
Boscovich, Roger, 72, 172, 197, 220
Boswell, James, 77
Bourguet, Louis, 84, 245
Bramhall, John, Bishop, 250
Briggs, William, 65
Browne, Peter, 249
Buffier, Claude, 55, 77
Buffon, Comte de, *see* Leclerc, Georges Louis
Burnet, Thomas, 6, 84, 245
Burnett, James, Lord Monboddo, 69
Bute, 3rd Earl of, *see* Stuart, John
Butler, Joseph, Bishop, 51, 71, 140

Campbell, George, 11
Campbell Fraser, A., 31, 67, 70
Cavendish, Henry, 29
Cavendish, William, Duke of Newcastle, 151
Clarke, Samuel, 14, 15, 32, 47, 51, 59, 75, 133, 231, 249
Collins, Anthony, 7, 32, 75, 133, 151, 231, 249
consubstantiation, 195
Cooper, Anthony Ashley, 3rd Earl of Shaftesbury, 140
Costa, Emanuel Mendes Da, 4
Cotes, Roger, 178, 253
  Cotes's theorem, 83
Creech, William, 35, 71
Croone, William, 20
Cudworth, Ralph, 162
Cullen, William, 65, 76

Daubenton, Louis Jean-Marie, 83
Deism, 32, 76
Desaguliers, J. T., 108, 248
Descartes, René, 7, 23, 24, 27, 65, 159, 174, 181, 187, 251
  theory of vortices, 60, 167, 168, 190
Ditton, Humphrey, 32
Drummond, Robert Hay, Archbishop of York, 55, 77
Du Hamel de Monceau, Henri-Louis, 3
Dumbarton Castle, 71, 76

Edinburgh, 7, 25, 27
  University of, 62
Edmond, Francis, 253
elephants, imaginary, 166, 175
emboîtement, 9, 14, 16
Enfield, William, 54, 76
Epicurus, 37, 132, 133, 140, 142
epigenesis, 12, 13
Euclid, 40, 71
Euler, Leonhard, 51, 75, 231

Fallopio, Gabriele, 60
fatalism, 17, 93, 127, 164
Ferguson, Adam, 41
final causes, 7

Index

Findlay, Robert, 38

Galen, 105, 248
Galilei, Galileo, 167, 216, 217
Garrick, David, 144, 250
Gay, John, Revd, 128, 136, 248, 250
generation
  equivocal, 11
  spontaneous, 11, 13
  theory of, 6–8, 9–19, 59, 60, 61, 62, 85–91, 93–8
George III, 54, 55
Glasgow, 4, 24, 31, 35, 40, 58, 60, 69
  Literary Society, x, 16, 17, 18, 27, 28, 37, 38, 39, 40, 47, 48, 70
  University of, 6, 7, 8, 9, 11, 26, 27, 28, 34, 56, 70, 166, 180
Gordon, Thomas, ix
Graff, Regnier de, 87, 246
Grant of Monymusk, Sir Archibald, 3
Gregory, James, 5, 71
Gregory, John, 49, 57, 65, 66, 74
Grew, Nehemiah, 4

Haller, Albrecht von, 27, 63, 64, 247
Hamilton, Sir William, 10
Hartley, David, 16, 25, 33, 34, 35, 36–8, 69, 93, 127, 128–60, 163, 168, 248, 249, 250
  Observations on man, 33, 34, 36, 69, 128–32, 134, 145, 146, 147, 248–9, 250
Hartsoeker, Nicolaas, 60
Harvey, William, 13, 60, 61, 87, 246
Hay, Thomas, 8th Earl of Kinnoul, 77
Helvétius, Claude Adrien, 8, 32
Hérissant, François-David, 98, 247
Hill, 'Sir' John, 4
Hobbes, Thomas, 37, 132, 133, 136, 142, 151, 250
Holbach, Paul Henri Thiry, Baron d', 68
Home, Henry, Lord Kames, 9, 10, 13, 14–15, 16, 18, 19, 33, 35, 42, 62
  The gentleman farmer, 13, 15, 62
Hooke, Robert, 11
Hume, David, 30, 34, 37, 71, 136, 138, 153, 166, 174, 175, 250
Hunter, John, 29, 49, 50
Hutcheson, Francis, 33, 140

ideas
  abstract, 37, 132, 142–5, 249
  association of, 35, 36, 37, 129, 130, 136–41, 144
  complex, 37, 142–5
  innate, 77
  of reflection, 37, 142–5, 166
  of sensation, 37, 166
  simple, 129, 144
instinct, 17, 37, 97, 109, 141–2

irritation/irritability, 17, 21, 22, 27, 49, 63, 93, 101–2, 227–9
Irvine, William, 60

Johnson, Joseph, 55
Julius Caesar, 134, 135
Jurin, James, 116, 248

Kames, Lord, *see* Home, Henry, Lord Kames
King, William, Archbishop, 128, 249
Kinnoul, 8th Earl of, *see* Hay, Thomas
Kunckel, Johann, 4

Laertius, Diogenes, 251
Langton, Bennet, 77
La Mettrie, Julien Offray de, 27, 49
Laudan, Larry, 33, 34
Lavoisier, Antoine-Laurent, 75
Leclerc, Georges Louis, Comte de Buffon, 4, 11, 12, 15, 19, 133
  his theory of generation, 6–8, 85–91
  his theory of the earth, 6, 58, 84–5
  Histoire naturelle, 4–9, 19, 83–92, 245, 246
  on mathematical truths, 5, 83
  on taxonomy, 5, 83
  on the nature of animals and humankind, 8–9, 91–2
Lee, Arthur, 78
Leeuwenhoek, Anthony van, 11, 60, 87, 246
Leibniz, G. W. F., 42, 85, 172, 245
Linnaeus, Carl, 4, 19, 83
Locke, John, 23, 24, 35, 51, 132, 136, 137, 139, 141, 142, 146, 158, 166, 167, 174–5, 180–2, 230–1, 249, 250
London, 55, 56, 65
London chronicle, 35
Lucretius, 164

McCosh, James, 31, 253
Malebranche, Nicholas, 23, 24, 29, 120, 142
Malpighi, Marcello, 4, 61, 87, 246
materialism, 6, 7, 8, 9, 11, 12, 13, 15, 18, 21, 27, 30–56, 58, 62, 67, 68, 75, 125–241
Maupertuis, Pierre-Louis Moreau de, 7, 246
mechanism, 8, 21, 25, 27, 29–30
Melvill, Thomas, 72
Monboddo, Lord, *see* Burnett, James
Monthly review, 36, 37, 38, 249
Murray, James, Lord Tullibardine, 248
muscular motion, 10, 20–2, 26–30, 101–24, 136, 141, 202–3

naturalism, 6, 58
necessitarianism, 30, 36, 37, 38, 150–1, 161–2
Needham, John Turberville, 7, 11, 12, 14, 88, 246
Newcastle, Duke of, *see* Cavendish, William
Newton, Sir Isaac, 15, 31, 65, 146, 147, 151,

Index

154, 161, 164, 172, 215–16, 217, 219, 220, 238, 239
on action at a distance, 46, 210–11
on attraction and repulsion, 167–8, 195–6, 203, 207, 213
on hypotheses, 45, 154, 178, 188, 209, 212, 213
and laws of motion, 46, 201
Letters to Bentley, 46, 73, 254
method of analysis and synthesis, 147
Optice, 214, 254
Opticks, 184, 188, 254–5: ether hypothesis, 45, 212–14, 220; Queries, 45
Principia, 5, 72, 152, 182–4, 188, 253: General Scholium, 251, 254, 255; Rules of philosophizing, 44, 71, 72, 161, 164, 165, 167, 168–9, 171, 172, 182–92, 196, 202, 203, 209, 210, 217, 230, 238, 252, 253, 254
theory of gravitation, 45, 184, 191, 208–9, 210, 220

occasionalism, 29–30, 42, 67, 120, 172–3
orang-utan, 8–9, 59, 133
Oswald, James, 32, 35, 53, 54, 55, 74
ovism, 10, 12

Parsons, James, 91, 246
Petty, William, Lord Shelburne, 54
Plato, 158, 159, 251
polyps, 13, 16, 18, 49, 95–6
Pope Leo X, 135
Porterfield, William, 25
pre-existence, theory of, 11, 13, 14, 16, 60, 61
preformationism, 11, 13, 60, 61
pre-ordained harmony, 42, 172
Price, Richard, 35, 38, 62, 69, 70, 73, 74, 76, 78, 161, 163, 166, 177
on inertia of matter, 74
on solidity and impenetrability of matter, 161
Priestley, Joseph, 16, 18, 30–56, 217, 230, 233, 238, 250
A free discussion, 38, 69, 74, 161–2, 171, 173, 176
Disquisitions, 16, 37, 38, 39, 42, 44, 45, 46, 52, 161, 162, 163–4, 168–71, 173, 175, 176, 182, 192, 208
Examination, 35, 43, 55, 56, 69, 74, 146, 163, 177, 178, 180
Hartley, 35–7, 51, 132–54, 162
Illustrations, 173, 204, 211, 214
Institutes, 32, 55, 69, 127–8, 146, 163, 177, 178, 180
politics of, 53–6
Socinianism, 47, 164
Pringle, Sir John, 29, 248

Ray, John, 4, 245

Réaumur, René Antoine Ferchault de, 3, 4, 11, 85, 97, 245
reductionism, 25
Reid, Thomas
Aberdeen Philosophical Society discourses, 40, 71
Active powers, 38, 40, 70, 71
and agricultural improvement, 3, 57
on attraction and repulsion, 44–5, 48, 165, 170, 171, 207–15, 219, 220–2, 239
on Bonnet, 16–18, 93–8
on Boscovich, 72, 172, 197, 220–1
on Buffon, 5–9, 14, 15, 19, 34, 35, 59, 83–92
correspondence with Lord Kames, 9, 13–14, 42, 61, 74
on existence of immaterial beings/causes, 3, 4, 7, 18, 21, 30, 31, 48, 50, 74–5, 217–19, 229–30, 232, 234–5, 239–40, 241
Glasgow Literary Society discourses, 16, 17, 18, 27, 28, 38–40, 48–52, 70, 71
on gravitation, 208–12, 239–40
on Hartley, 33, 34, 35, 36–8, 69, 70, 128–32, 145–54, 154–60
on hypotheses, 6, 24–5, 28, 33–6, 42–3, 70, 118–19, 150–4, 156–8, 184, 187–8
Inquiry, 9, 24, 25, 33, 34, 38, 40, 43, 55, 68, 70, 71, 248, 251
Intellectual powers, 9, 33, 34, 37–8, 40, 70, 71, 251, 252
lectures: Glasgow, 6, 7, 8, 9, 12, 15, 24, 26, 27, 28, 34, 40, 56; King's College Aberdeen, 3–4, 10–11, 19, 21, 27, 34, 56, 57, 63, 70, 71, 72, 73
on laws of motion, 204–6
on love of simplicity, 43
on matter: animated, 48, 223–4; existence of, 166, 176–80; inanimate, 48, 218–23, 232; inertia of, 44, 167, 169, 171, 172, 191, 201–7, 216, 217, 222, 230, 231–2, 233; organization of, 7, 49, 74, 225–6, 240; passivity of, 7, 17, 20–1, 22, 48, 73, 217, 218, 222, 230, 233, 238–9; solidity and impenetrability of, 46, 73, 164–5, 167, 169–71, 173, 192–200
minister at New Machar, 3
on muscular motion, 20–2, 26–30, 40, 101–24, 202–3
as natural historian, 19–20
on nervous power, 28–9, 118–24
on nescience, 12, 17, 18, 19, 26, 27, 30, 49, 221–2, 227–9, 233–4, 236
Newtonianism of, 5, 31, 47, 67: on Newton's ether hypothesis, 42, 45, 73, 212–14, 220; on Newton's rules of philosophizing, 30, 34, 38, 41–5, 71, 72, 73, 161, 167, 168–9, 171, 172, 182–92, 202, 203, 206–7, 209, 215–16, 217
orations, 34, 71

on physiology of perception, 22–5, 30, 34, 114–17, 123–4, 159–60
on political utopias, 40
politics of, 53
on Priestley, 30–56, 127–8, 132–49, 161–71, 173–217
on principle of life, 18, 27, 48–50, 224–7, 236–7, 240
on relation between mind and body, 26, 29, 30, 51–2, 66, 235
on scope of natural philosophy, 41, 42
'Some Observations on the Modern System of Materialism', 31, 39, 40, 41–8, 52–3, 173–217, 253, 266–7
on Stahl, 26, 27
on Stay, 171–3
on theory of generation, 9–19
on theory of ideas, 22–4, 34, 158–60
theory of organized atoms of, 9–10, 13–17, 19, 49
on vibrations and vibratiuncles, 129, 135, 138, 149, 154–8, 168
on Whytt, 21–2, 27, 30, 101–3
Roscius, Quintus, 144, 250
Rose, Samuel, 70
Rose, William, 36, 70
Royal College of Physicians (London), 20
Royal Society of London, 20, 75

Scheutzer, Jacob, 85, 245
Scott, Robert Eden, ix
Shaftesbury, 3rd Earl of, *see* Cooper, Anthony Ashley
Shelburne, Lord, *see* Petty, William
Skene, David, 4, 11, 12, 19, 20, 57
Skene, George, 4, 57
Smellie, William, 59
Smith, Robert, 248
Spallanzani, Lazzaro, 12, 97, 247

St. John, Henry, Lord Bolingbroke, 32, 76, 151
Stahl, G. F., 26, 27, 49, 50
Stay, Benedetto, 72, 171–3, 252–3
Steno, Nicolaus, 85, 245
Stewart, Dugald, 3, 31, 67
Stewart, John (Aberdeen professor), 248
Stewart, John (Edinburgh professor), 7, 62
stimulus, 17, 22, 27, 49, 64, 93, 101–3, 227–9
Stuart, Alexander, 65, 248
Stuart, John, 3rd Earl of Bute, 54, 55
Swammerdam, Jan, 11

Thirty-nine Articles, 55, 56
Toleration Act (1689), 55
Tournefort, Joseph Pitton de, 3, 4
transubstantiation, 195
Trembley, Abraham, 13, 49
trinitarianism, 47, 194
Tullibardine, Lord, *see* Murray, James
Turnbull, George, 75
Tytler, Alexander Fraser, Lord Woodhouselee, 9–10, 13, 16, 17

Vallisneri, Antonio, 87, 246

Walker, John, 15, 62
Warrington Academy, 54, 76
Whiston, William, 6, 84, 245
Whytt, Robert, 21–2, 25, 27, 28, 49, 50, 64, 67, 74, 75, 247
Wilkins, John, 131, 249
Woodhouselee, Lord, *see* Tytler, Alexander Fraser
Woodward, John, 6, 84, 245

xenophobia, English, 54–5, 77, 78

Yolton, John, 31, 68